SCIENCE PLUS®

TECHNOLOGY AND SOCIETY

LEVEL BLUE

TEST GENERATOR
TEST ITEM LISTING

HOLT, RINEHART AND WINSTON

Harcourt Brace & Company

Austin • New York • Orlando • Atlanta • San Francisco • Boston • Dallas • Toronto • London

Photo/Art Credits
Abbreviated as follows: (t) top; (b) bottom; (l) left; (r) right; (c) center; (bkgd) background.
Front Cover: (bkgd), Page Overtures; (bl), Randy Gates/Morgan Cain & Associates. Back cover: (bkgd), Page Overtures; (bl), Jeff Smith/FotoSmith/Reptile Solutions of Tucson. Title Page (i): (bkgd), Page Overtures; (bl), Jeff Smith/FotoSmith/Reptile Solutions of Tucson.

All Art and Photos, unless otherwise noted, are contributed by Holt, Rinehart and Winston.

SCIENCEPLUS is a registered trademark of Harcourt Brace & Company licensed to Holt, Rinehart and Winston, Inc.

Printed in the United States of America

ISBN 0-03-095794-X   3 4 5 6 7 8 9 021 99

Contents

	Page

Contents, continued

Page

Using the *SciencePlus Test Generator*

The *SciencePlus Test Generator* contains assessment items for each unit, chapter, and SourceBook unit of your textbook. For your easy reference, this *Test Item Listing* contains a complete printout of the questions and answers that come with the *SciencePlus Test Generator.* It also contains information that is specific to the installation and use of the program with *SciencePlus.*

The *User's Guide,* which is included in the *Test Generator* package, contains general directions for using the program.

Test Generator Diskettes

The *Test Generator* is delivered on a series of diskettes. The Program Disk or Disks contain the *Test Generator* program, and the Test Items Disks contain the assessment items.

Question Types

The test items are classified by question type. In the *Test Generator,* question type is called an attribute. The *Test Generator* lets you specify attributes when selecting questions for a test. It also allows you to view and to edit question attributes. Your *User's Guide* describes these features of the program.

The *SciencePlus Test Generator* has thirteen types of questions stored on the Test Items Disks. The question types are identified by a one-letter abbreviation, as shown in the chart at right.

Question Difficulty Level

Some of the chapter assessment items and end-of-unit assessment items are meant to be more challenging. The *Test Generator* has an attribute for level of difficulty. Your *User's Guide* describes how to use the difficulty levels when selecting questions for a test. In *SciencePlus,* challenging questions are identified by the following one-digit level of difficulty:

3   Difficult
4   Very Difficult

Question Types and Codes	
A	Activity Assessment
C	Correction/Completion
D	Data for Interpretation
E	Short Essay
G	Graphic
I	Illustrative
M	Multiple Choice
N	Numerical Problem
P	Performance Task
S	Short Response
T	True/False
W	Word Usage
K	Matching

Macintosh® Users

The *Test Generator User's Guide* describes how to copy the diskettes to your computer's hard drive. Look for the section on installation.

Question Disks: Folders and Files

On the Test Items disks, the files are grouped into folders, with one folder for each unit of your textbook. Each folder contains one chapter file for each chapter in the unit and four additional files. The additional files are a unit test, a SourceBook test, an activity assessment, and a bank of extra assessment items for the unit.

For example, here is a breakdown of the files in the Unit 1 folder.

Chapter.C1	Chapter 1 Assessment
Chapter.C2	Chapter 2 Assessment
Chapter.C3	Chapter 3 Assessment
Activity.A1	Activity Assessment for Unit 1
Extra.E1	Extra Assessment Items for Unit 1
Source.S1	SourceBook Assessment for Unit 1
Unit.U1	End-of-Unit Assessment for Unit 1

As the *User's Guide* explains, you can make tests by combining questions from any of the files.

Windows® Users

The *Test Generator User's Guide* describes how to install the diskettes on your computer's hard drive. Look for the section on installation.

Question Disks: Files and Directories

The test questions are stored in chapter files. After you have installed the *Test Generator* on your hard drive, the chapter files are grouped into directories, with one directory for each unit of your textbook. Each directory contains one chapter file for each chapter in the unit and four additional files. The additional files are a unit test, a SourceBook test, an activity assessment, and a bank of extra assessment items for the unit.

For example, here is a breakdown of the files in the Unit_1 directory.

Chapter.C1	Chapter 1 Assessment
Chapter.C2	Chapter 2 Assessment
Chapter.C3	Chapter 3 Assessment
Chapter.A1	Activity Assessment for Unit 1
Chapter.E1	Extra Assessment Items for Unit 1
Chapter.S1	SourceBook Assessment for Unit 1
Chapter.U1	End-of-Unit Assessment for Unit 1

As the *User's Guide* explains, you can make tests by combining questions from any of the files.

Technical Support

If you have any questions about the *Test Generator* or need assistance, call the Holt, Rinehart and Winston technical support line at 1-800-323-9239.

About *SciencePlus* Assessment

SciencePlus assessment is designed to provide you with a comprehensive set of evaluation tools. For each unit of the Pupil's Edition, you will find five categories of tests: Chapter Assessment, End-of-Unit Assessment, SourceBook Assessment, Activity Assessment, and Extra Assessment Items. The Extra Assessment Items consist of questions found only in the *Test Generator* and *Test Item Listing*. All of the other tests have blackline-master versions in the *Teaching Resources* booklets.

This *Test Item Listing* booklet provides a handy way of previewing the tests and questions contained on the *Test Generator*. The *Test Generator* is easy to use, includes graphics, and is fully customizable. You can easily change questions or add questions of your own.

Each test is self-contained and comprehensive. You will find that the Chapter Assessment and End-of-Unit Assessment each contain at least two Challenge questions to help you evaluate higher-order thinking skills. Activity Assessment, SourceBook Assessment, and Extra Assessment Items are also included for each unit.

The Activity Assessments help you assess the way students actually carry out Explorations and experiments. These assessments require students to manipulate equipment and materials and to compile and analyze data. Using the Activity Assessments, you can assess your students' abilities to solve problems using the tools, equipment, and techniques of science. Teacher's Notes are provided for each Activity Assessment to guide you in setting up the activity stations and in grading the students' efforts.

A Note About the Philosophy

In *SciencePlus* there is no clear distinction between teaching and assessing. Every suggested assessment activity, including testing, is designed to teach. Assessment should be ongoing and should measure performance in every area. The quality of student class work, homework, lab work, ScienceLog entries, and performance on tests should all be factors in assigning grades.

The authors strongly discourage reliance on recall-based assessment items. The *SciencePlus* program relies on performance-based assessment strategies. Students are provided with realistic problems to solve. Much effort has been directed toward making the assessment items consistent with the actual pursuit of scientific knowledge. Science is not a process of memorizing vocabulary; science consists of gathering data, hypothesizing, experimenting, making predictions, analyzing data, making generalizations, and many other processes. The test items in this program are designed to assess your students' mastery of these skills. In working toward this goal, the authors have used the Assessment Item Development model shown on the following page.

Assessment Item Development Model

Verbal

Word Usage: Words given are to be used in a prescribed situation.

Correction/Completion: Incorrect or incomplete sentences and paragraphs are given for correction or completion.

Short Essay: Information is given or a question is posed for short-essay response.

Short Response: Answers to these questions require a tick mark, a line, a single word, a phrase, or a sentence.

Graphic

Graph for Interpretation: A graph of a relationship between two variables is given for interpretation.

Graph for Correction or Completion: An incorrect or incomplete graph is given for correction or completion.

Graphing Data: Data are given to be graphed.

Illustrative

Illustration for Interpretation: Illustrations (drawings or photographs) are presented for interpretation.

Illustrations for Correction or Completion: An incorrect or incomplete illustration is given for correction or completion.

Answering by Illustration: A question is asked for which a drawing is the expected answer.

Numeric

Data for Interpretation: A data table is given for interpretation.

Numerical Problem: A problem is given for a numerical solution.

Selecting and Using the Assessment Items

When choosing assessment items that are appropriate for your class, consider the students' reading and writing levels, process skills, and critical-thinking skills. A brief glance at the questions, illustrations, and graphs in each test will give you a good idea of the most suitable questions for your students.

Keep in mind that the manner of testing determines what is learned: tests that require tick-mark responses teach tick-marking. Tests that require verbal, graphic, illustrative, and numeric responses develop writing, speaking, graphing, drawing, and mathematical skills. A superior test draws on as many skills as possible.

In addition to the other tests, each Activity Assessment is designed to be done in one class period. Of course, the Activity Assessments will require some advance preparation. The Teacher's Notes provide clear directions for setting up each assessment. Every effort has been made to make the Activity Assessments economical in terms of both time and materials.

Chapter 1 Assessment

Word Usage

1. For each group of words below, write one or two sentences to show how the words are related.

   a. carbon dioxide, oxygen, photosynthesis, green plants

   _____

   _____

   _____

   b. sunlight, water, carbon dioxide, starch, green plants, leaves

   _____

   _____

   _____

Short Response

2. Write a word equation to represent the process of photosynthesis. (If you need a hint, see the words in the question above.)

   _____

   _____

   _____

   _____

3. If a plant leaf is compared with a factory, what could each of the following represent?
   a. the factory's energy source

   _____

   b. the factory's raw materials

   _____

   c. the factory's finished product

   _____

Graphic

4. Use the following graph to answer the questions below.
   a. Explain why most tree leaves appear to be green and not blue.

   _____

   _____

   _____

   _____

   b. What other conclusions can you draw from the data in the graph?

   _____

   _____

   _____

   _____

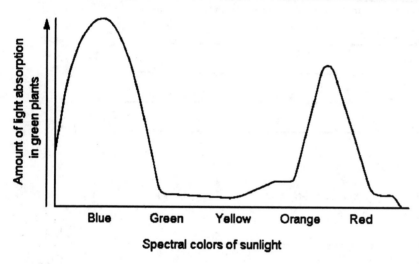

Short Essay

5. Owls eat small animals, such as snakes and mice, but do not eat plants. If the Earth experienced a drastic change such as a prolonged lack of sunlight, explain how the owls might be affected.

   _____

   _____

   _____

   _____

   _____

   _____

Answers to Chapter 1 Assessment

Word Usage

1. Answer:
   a. Sample answer: During *photosynthesis*, *green plants* take in *carbon dioxide* and release *oxygen*.
   b. Sample answer: In the presence of *sunlight*, *green plants* use *carbon dioxide* and *water* to manufacture *starch*, which they store in their *leaves*.

Short Response

2. Answer:
   Sample answer: Carbon dioxide + water (in the presence of sunlight and chlorophyll) → starch + oxygen

3. Answer:
   a. Sunlight
   b. Water and carbon dioxide
   c. Starch (or glucose) and oxygen

Graphic

4. Answer:
   a. Sample answer: Most trees are green plants whose leaves absorb blue light, among other colors of light, and reflect green light. We see the color that is reflected, not the color that is absorbed.
   b. Answers will vary but should be consistent with the data in the graph and with the content of the unit. (For example, plants would probably grow better in blue light than in yellow light.)

   Difficulty: 3

Short Essay

5. Answer:
   Sample answer: Even though owls do not eat plants directly, they eat animals, such as mice, that eat plants, as well as animals, such as snakes, that eat plant-eating animals, like mice. If there were no sunlight, photosynthesis could not take place, and plants could neither produce food in order to grow nor furnish food for plant-eating animals. The owls would therefore perish if photosynthesis were not possible.

   Difficulty: 4

Chapter 2 Assessment

Word Usage

1. Use all of the following terms in one or two sentences to show how the terms are related: *osmosis, particles, concentration,* and *membrane.*

   _____

   _____

   _____

   _____

   _____

Correction/Completion

2. The statements below are incorrect or incomplete. Your challenge is to make them correct and complete.

   a. The paper filter for a coffee maker should be permeable to both water and coffee grounds.

      _____

      _____

      _____

   b. A semipermeable membrane will allow a certain amount of any substance, regardless of particle size, to pass through it.

      _____

      _____

      _____

Short Response

3. During a sunny day, a green plant takes in carbon dioxide and releases oxygen.

   a. What gas does the green plant take in at night?

   _____

   b. What gas does the green plant release at night?

   _____

   c. What else does the green plant release constantly?

   _____

Illustrative

4. Draw a diagram to show how water from the ground is transported to the leaves of a plant.

5. Nick's grandmother added two spoonfuls of sugar to a cup of very hot tea, but before she could stir it, she was called away for 5 minutes. Make a diagram to show what the cup of tea with sugar would look like, on the particle level, after 5 minutes. Make a key that shows the difference between the tea particles and the sugar particles.

Short Essay

6. According to legend, after the Roman army defeated the people of Carthage, a city in North Africa, the Roman conquerors not only burned all of the buildings but also spread salt over the farmland. What was the purpose of doing this, and why would it be effective?

   _____

   _____

   _____

   _____

   _____

   _____

   _____

   _____

   _____

   _____

Answers to Chapter 2 Assessment

Word Usage

1. Answer:
   Sample answer: During the process of *osmosis, particles* of water or some other
   substance flow through a *membrane* until the *concentration* of particles is equal on both
   sides of the membrane.

Correction/Completion

2. Answer:
   a. The paper filter for a coffee maker should be permeable to water *but impermeable to
      coffee grounds.*
   b. A semipermeable membrane will allow *small particles, but not large ones,* to pass
      through it.

Short Response

3. Answer:
   a. Oxygen
   b. Carbon dioxide
   c. Water

Illustrative

4. Answer: Sample diagram:

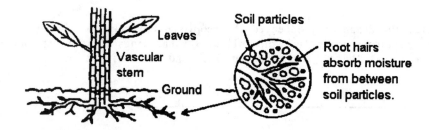

Difficulty: 3

5. Answer: Sample diagram:

Difficulty: 3

Short Essay

6. Answer:

Sample answer: By salting the fields, the Romans hoped to ensure that crops would not grow, thus preventing the people of Carthage from ever rebuilding their city. It would work because most crops cannot grow in a salty environment. The plants (or seeds) need to take moisture out of the soil, but if the soil is salty, it will draw fresh water out of the plants (or seeds) and will not let moisture into the plants (or seeds).

Difficulty: 4

Chapter 3 Assessment

Word Usage

1. Use the terms *oxygen* and *carbon dioxide* to explain how plants and animals depend on each other to carry out life processes.

_____

_____

_____

_____

_____

_____

Correction/Completion

2. Complete the following sentences:

   a. Thyroxin, which is produced by the _____ gland, regulates the rate at which _____ is used by the body.

   b. The pancreas produces the chemical insulin, which allows _____ to be used in the process of _____.

Short Response

3. Label each list with the name of the appropriate human body system.

a. _____	b. _____	c. _____	d. _____
teeth	brain	heart	nose
stomach	neurons	capillaries	lungs
intestines	synapses	arteries	blood

4. The diagram below shows the blood vessels in a fish's gills. Add arrows to the close-up view to show the path that the blood takes through the gills. Label the two largest vessels as carrying either oxygen-rich blood or oxygen-poor blood.

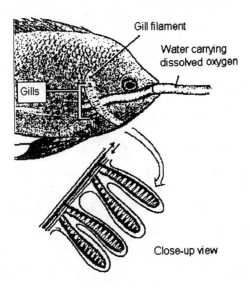

Gill filament

Water carrying dissolved oxygen

Gills

Close-up view

Short Essay

5. How does a normal concentration of carbon dioxide in the atmosphere benefit the Earth, and what might happen if there were too much carbon dioxide?

_____

_____

_____

_____

_____

_____

_____

_____

_____

6. Why might it be better for astronauts to take plants rather than animals on long space missions?

_____

_____

_____

_____

_____

_____

_____

_____

Answers to Chapter 3 Assessment

Word Usage

1. Answer:
   Sample answer: Plants conduct photosynthesis and grow by taking in *carbon dioxide* and giving off *oxygen*, while animals take in oxygen and give off carbon dioxide in the process of respiration. Without plants, animals would run out of oxygen, and without animals, plants would not have enough carbon dioxide.

Correction/Completion

2. Answer:
   a. Thyroxin, which is produced by the <u>thyroid</u> gland, regulates the

   rate at which <u>food</u> is used by the body.

   b. The pancreas produces the chemical insulin, which allows <u>sugar</u>

   to be used in the process of <u>respiration</u>.

Short Response

3. Answer:

a. *Digestive*	b. *Nervous*	c. *Circulatory*	d. *Respiratory*
teeth	brain	heart	nose
stomach	neurons	capillaries	lungs
intestines	synapses	arteries	blood

Illustrative

4. Answer:

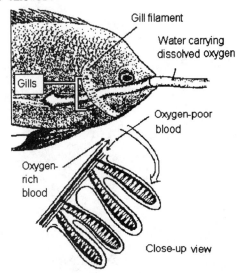

Gill filament

Water carrying dissolved oxygen

Gills

Oxygen-poor blood

Oxygen-rich blood

Close-up view

5. Answer:
Sample answer: Carbon dioxide prevents the Earth from becoming too cold because carbon dioxide traps some solar energy on the Earth during the day and prevents some low-level energy from leaving the Earth at night. If carbon dioxide levels were to increase, too much energy would be prevented from leaving the Earth at night, and the atmosphere could heat up. This greenhouse effect could change the Earth's climate.

Difficulty: 3

6. Answer:
Sample answer: Plants "recycle" carbon dioxide by turning it into oxygen, which the astronauts can breathe. In addition, plants manufacture their own food, which the astronauts could eat, and plants produce very little waste material. Animals, on the other hand, need oxygen and food, just as the astronauts do, and animals produce wastes.

Difficulty: 4

Unit 1 End-of-Unit Assessment

Word Usage

1. For each group of words below, write a sentence to show how the words are related.

   a. green plants, photosynthesis, sunlight

   _____

   _____

   _____

   b. water particles, membrane, root hairs

   _____

   _____

   _____

Correction/Completion

2. Correct the following paragraph:

   Shirley discovered that water will move through a semipermeable membrane into a concentrated sugar solution. This suggests that osmosis involves the movement of water particles from an area of low concentration to an area of high concentration.

   _____

   _____

   _____

   _____

   _____

   _____

Short Response

3. How might you demonstrate that water passes through a plant into the air?

_____

_____

_____

_____

_____

_____

4. Chemical reactions involve raw materials (called reactants), products, and energy. Consider the chemical processes below and complete the table.

Process	Raw materials	Direct energy source (including stored energy)	Products
photosynthesis in an African violet			
respiration in a human			
respiration in a beech tree			

5. Suppose that you watered your plant with sea water rather than fresh water. What would happen? Why?

_____

_____

_____

_____

_____

_____

_____

Illustrative

6. The labels on the diagram below are incorrect. Rearrange the labels so that they are in the correct location for each process.

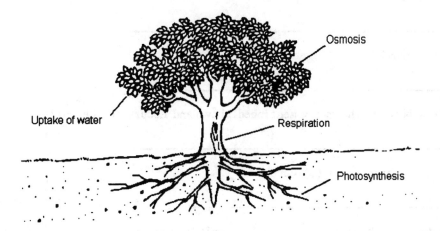

Osmosis

Uptake of water

Respiration

Photosynthesis

Graphic

7. The graph shows the relationship of temperature to the rate of photosynthesis. Use the graph to answer the questions that follow.

a. Is the best temperature for photosynthesis cold, moderate, or hot?

_____

b. Why do you think this relationship exists between the rate of photosynthesis and temperature?

_____

_____

_____

_____

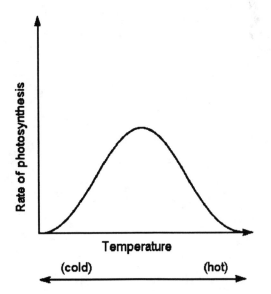

Illustrative

8. Use the diagram below to answer the following questions.

a. What gas does the cow exhale?

_____

b. What gas does the cow inhale?

_____

c. What process involves both of the gases mentioned above and occurs in the cow?

_____

d. Where in plants does the process in (c) occur?

_____

e. Name the process involving these gases that occurs only in plants.

_____

f. Why are the interactions depicted in the diagram called a cycle?

_____

_____

_____

_____

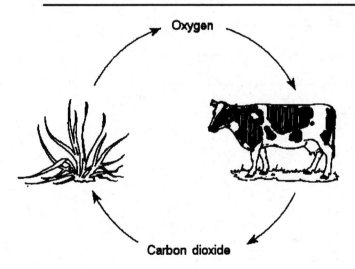

Oxygen

Carbon dioxide

Short Response

9. A cell membrane, such as the membrane of an egg, is often compared with a screen or sieve. Use this comparison to explain why water travels through a membrane into an egg, while the materials inside the egg do not travel out. Illustrate your answer in the space provided below.

_____

_____

_____

_____

_____

_____

Graphic

10. The graph below shows the results of a study on the factors that affect the amount of food made by photosynthesis in a cucumber leaf. Use it to help you answer the questions that follow.

a. What was measured to show the amount of photosynthesis that took place?

_____

_____

b. This study tested the effects of three variables. What were they?

_____

_____

c. What combination of factors resulted in the highest rate of photosynthesis?

_____

_____

d. What factor caused the greatest increase in the amount of photosynthesis?

_____

_____

e. Which factor, if any, seems to be the most important in increasing the rate of food production? How can you tell?

_____

_____

_____

_____

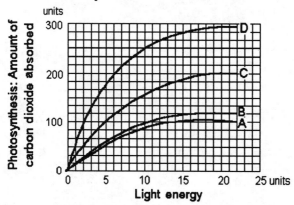

Photosynthesis in a Cucumber Leaf

Percentage of Carbon Dioxide in Atmosphere	
A 0.03% carbon dioxide at 20°C	B 0.03% carbon dioxide at 30°C
C 0.13% carbon dioxide at 20°C	D 0.13% carbon dioxide at 30°C

Correction/Completion

11. Complete the paragraph below using words from the following list: *bloodstream*, *cell membranes*, *diffusion*, *intestinal wall*, *mouth*, *small intestine*, *starch*, and *sugar*.

Digestion and respiration work together to convert food into fuel. The conversion process starts in the _____, where saliva starts to change _____ into _____. The process continues in the stomach and _____, where food is mashed up and chemically broken down into water-soluble particles small enough to penetrate the _____ and enter the _____. The blood carries the dissolved nutrients throughout the entire body. _____ carries the dissolved nutrients through the _____ into the individual cells.

Short Essay

12. Explain two important reasons why plants are essential to human life.

_____

_____

_____

_____

_____

_____

_____

_____

_____

_____

_____

Answers to Unit 1 End-of-Unit Assessment

Word Usage

1. Answer:
   a. Sample answer: *Green plants* use the energy of *sunlight* to make starch during *photosynthesis*.
   b. Sample answer: *Water particles* move into a plant's cells through the cell *membranes* of the plant's *root hairs*.

Correction/Completion

2. Answer:
   Sample correction: Shirley discovered that water will move through a semipermeable membrane into a concentrated sugar solution. This suggests that osmosis involves the movement of water particles from an area of *high* concentration to an area of *low* concentration.

Short Response

3. Answer:
   Sample answer: Place a clear plastic bag over a growing plant, secure it with a rubber band, and leave it there for a period of time, perhaps as long as a week. Water will collect on the inside of the bag. This shows that the water is the result of transpiration, or the passing of water from the plant into the air.

4. Answer:

Process	Raw materials	Direct energy source (including stored energy)	Products
photosynthesis in an African violet	carbon dioxide, water	sunlight	sugars (or starch), oxygen
Respiration in a human	oxygen	food	carbon dioxide, water
Respiration in a beech tree	oxygen	sugars and starches made in photosynthesis and stored	carbon dioxide, water

5. Answer:
   The plant would wither and die because, during osmosis, water passes from a region of a higher concentration (within the plant's cells) to a region of lower concentration (the sea water). Without sufficient water, the plant's cells would shrink, and the plant would wither. Eventually, the plant would die.

Illustrative

6. Answer:

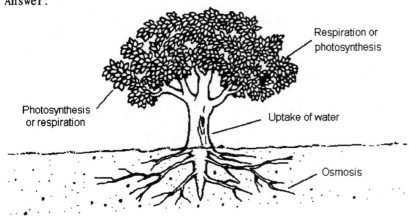

Respiration or photosynthesis

Photosynthesis or respiration

Uptake of water

Osmosis

Graphic

7. Answer:
   a. Moderate
   b. Sample answer: Very high or low temperatures make the raw materials of photosynthesis less available to the plant. Extreme temperatures can damage the parts of the plant that perform photosynthesis.

Illustrative

8. Answer:
   a. Carbon dioxide
   b. Oxygen
   c. Respiration
   d. Respiration occurs in all plant cells.
   e. Photosynthesis
   f. Sample answer: The interactions are called a cycle because the gases are endlessly transformed into one another by plants and animals.

Short Response

9. Answer:
   A cell membrane prevents the passage of the large egg particles (molecules) while permitting the passage of the smaller water particles (molecules).
   Illustrations will vary but should clearly and logically demonstrate the above process.

   Difficulty: 3

Graphic

10. Answer:
    a. Amount of carbon dioxide absorbed
    b. Light energy, carbon dioxide, temperature
    c. 20-25 units of light energy, 30°C, 0.13 percent carbon dioxide
    d. Percentage of carbon dioxide
    e. Sample answer: Percentage of carbon dioxide seems most important, according to the information in the graph, because the difference between lines B and D is greater than the difference between lines C and D.

    Difficulty: 3

Correction/Completion

11. Answer:

Digestion and respiration work together to convert food into fuel. The conversion process starts in the __mouth__ , where saliva starts to change __starch__ into __sugar__ . The process continues in the stomach and __small intestine__ where food is mashed up and chemically broken down into water-soluble particles small enough to penetrate the __intestinal wall__ and enter the __bloodstream__ . The blood carries the dissolved nutrients throughout the entire body. __Diffusion__ carries the dissolved nutrients through the __cell membranes__ into the individual cells.

Difficulty: 4

Short Essay

12. Answer:

Sample answer: Plants play two essential roles in sustaining human life. First, they remove carbon dioxide from the air, exchanging it for oxygen. Oxygen is essential to human life, while excess carbon dioxide is potentially harmful because of its role in the greenhouse effect (trapping heat and consequently changing the climate). Second (but equally important), plants store energy from the sun through photosynthesis. Animals can obtain some of this energy by eating plants or by eating animals that eat plants. Humans participate in this food chain as well and thus receive the benefits of the solar energy stored by the plants.

Difficulty: 4

Unit 1 Activity Assessment

Activity Assessment

1. **Burning Up**
   Teacher's Notes

   ### Overview
   Students burn several samples of food to observe the end products of respiration and then compare the processes of respiration and photosynthesis. They compile their findings in a Data Chart and use a diagram to summarize their findings.

   ### Materials
   (per activity station)
   - a ceramic or other type of fireproof dish
   - matches
   - tweezers
   - 2 test tubes with corks
   - 2 food samples to test, such as a sugar cube and peanut
   - 2 strips of blue cobalt chloride paper
   - 10 mL of limewater
   - safety goggles

   ### Preparation
   Prior to the assessment, equip student activity stations with the materials needed for each experiment. Remind students to exercise care when working with a flame.

   ### Time Required
   Each student should have 20 minutes at the activity station and 20 minutes to complete the Data Chart and diagram.

   ### Performance
   At the end of the assessment, students should turn in the following:
   - a completed Data Chart and diagram
   - a summary of their observations

   ### Evaluation
   The following is a recommended breakdown for evaluation of this Activity Assessment:
   - 30% appropriate and logical use of materials and equipment
   - 35% ability to make observations and correctly interpret results
   - 35% ability to compare two processes with a diagram

   ### Safety Alert!

2. Burning Up

You have observed that plants use carbon dioxide, water, and energy from sunlight to produce food and oxygen. This is the process of photosynthesis. You have also observed that animals and plants give off carbon dioxide, water, and energy. These substances are the products of respiration. Respiration is the chemical breakdown of food in the presence of oxygen. Although we cannot observe respiration directly, we can observe the breakdown of food into carbon dioxide, water, and energy in a similar process— burning. In the activity below, you will burn two samples of foods and observe the products. Then you will show the relationships that exist between the processes of photosynthesis and respiration in plants and animals.

Before You Begin . . .

As you work through the tasks, keep in mind that your teacher will be observing the following:
- how you use the materials to perform the experiment
- how well you make observations and correctly interpret results
- how you compare two processes with a diagram

Burn 'Em Up!

Task 1: Place one small sample of food in a fireproof dish, and carefully light it with a match. When the sample is burning well, pick it up with the tweezers and place it into a dry test tube. **Be Careful: Exercise extreme caution when moving a burning object.** Seal the tube with a cork, and watch what happens. Record your observations in your Data Chart. Then remove the sample and quickly put a piece of blue cobalt chloride paper into the test tube. Record your observations. Next, add a little limewater. Replace the cork and gently shake the tube. What happens?

Task 2: Using a different test tube, repeat Task 1 for the other sample. Record your observations in your Data Chart.

Task 3: Blue cobalt chloride paper turns pink in the presence of water. Limewater becomes cloudy in the presence of carbon dioxide. Summarize your observations and include a comparison of the substances produced in respiration and burning. Also comment on the relationship between respiration and photosynthesis.

Task 4: Finally, draw a diagram showing how photosynthesis and respiration are related. Be sure to include the energy from the sun, an animal, a plant, carbon dioxide, food, water, and oxygen.

Safety Alert!

3.

Data Chart

Sample	Observations		
	Burning	Cobalt chloride paper	Limewater
1			
2			

Summary of Observations

_____

_____

_____

_____

_____

_____

Diagram

Answers to Unit 1 Activity Assessment

Activity Assessment

1. Answer: Not applicable (teacher's notes)

2. Answer: Not applicable (student's notes)

3. Answer:

Data Chart

Sample	Observations		
	Burning	Cobalt chloride paper	Limewater
1	Answers will vary depending on types of food samples used, but students should see droplets of water.	The paper should turn pink because water is present.	The limewater should become cloudy because carbon dioxide is present.
2	Answers will vary depending on types of food samples used, but students should see droplets of water.	The paper should turn pink because water is present.	The limewater should become cloudy because carbon dioxide is present.

Summary of Observations

The fact that water and carbon dioxide were the products of the tests on the food samples shows that burning is a type of respiration. This activity gives some insight into the digestion and respiration of animals as well as the respiration of plants. The products of these tests also show how respiration contributes to the process of photosynthesis.

Diagram
Sample diagram:

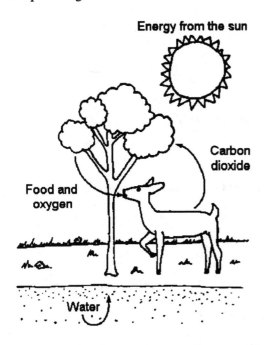

Unit 1 SourceBook Assessment

Multiple Choice

1. The element that forms the "backbone" of the molecules of life is

    a. carbon.   b. hydrogen.   c. oxygen.   d. nitrogen.

2. Ninety-nine percent of all living matter on Earth is made from _____ of the 92 naturally occurring elements.

    a. 1   b. 4   c. 16   d. 25

3. All organic compounds contain the element

    a. carbon.   b. hydrogen.   c. oxygen.   d. nitrogen.

4. Which of the following is NOT an organic compound that occurs in living things?

    a. carbohydrates   b. water   c. lipids   d. nucleic acid

5. A simple sugar is a

    a. monosaccharide.
    b. disaccharide.
    c. polysaccharide.
    d. pentasaccharide.

6. Amino acids are the building blocks of

    a. carbohydrates.   b. lipids.   c. proteins.   d. nucleic acids.

True/False

7. The only thing that cholesterol is known to do in animals is cause heart disease.

    a. true   b. false

Multiple Choice

8. The most common of the organic molecules that are found in living things are called

    a. carbohydrates.   b. lipids.   c. proteins.   d. nucleic acids.

9. When water molecules stick to materials that they cannot dissolve, they are demonstrating the property of

    a. adhesion.   b. capillary action.   c. cohesion.   d. heat capacity.

10. A condensation reaction, in which a disaccharide is produced from two monosaccharides, would also produce

   a. a protein.   b. water.   c. a lipid.   d. a carbohydrate.

True/False

11. Organisms that produce their own food are known as consumers.

   a. true   b. false

12. Osmosis is a type of diffusion.

   a. true   b. false

Multiple Choice

13. Osmosis involves the movement of

   a. salt molecules.
   b. sugar molecules.
   c. water molecules.
   d. any negative ion.

14. A fish swimming upstream is most like

   a. passive transport.
   b. active transport.

True/False

15. Glucose is the only fuel used by the body in respiration.

   a. true
   b. false

Multiple Choice

16. Photosynthesis occurs inside plant organelles called

   a. carbohydrates.
   b. chloroplasts.
   c. mitochondria.
   d. stomata.

17. Animals, plants, and many other organisms use _____ in cellular respiration.

   a. carbohydrates
   b. chloroplasts
   c. mitochondria
   d. glucose

18. Collectively, an organism's processes of energy conversion and of breaking down and producing chemicals are called

    a. synthesis.
    b. excretion.
    c. active transport.
    d. metabolism.

19. Which of the following is *not* an organ of excretion in the human body?

    a. lungs
    b. heart
    c. skin
    d. kidneys

20. Hormones are most involved with

    a. excretion.
    b. metabolism.
    c. regulation.
    d. obtaining raw materials.

Short Response

21. Using your knowledge of how atoms form bonds with other atoms, explain why an atom of carbon has the ability to bond with up to four different atoms at once.

    _____

    _____

    _____

    _____

    _____

    _____

22. Explain how the terms *monomer* and *polymer* are related to organic compounds.

    _____

    _____

    _____

    _____

    _____

    _____

23. List three food sources that contain complex carbohydrates.

    _____

    _____

    _____

24. Which of the four organic compounds that occur in living organisms would a marathon runner need most while competing? Explain your answer.

_____

_____

_____

_____

25. Write a statement comparing and contrasting DNA (deoxyribonucleic acid) and RNA (ribonucleic acid).

_____

_____

_____

_____

_____

_____

_____

26. One of the many functions of water, especially in living things, is to act as a temperature regulator that keeps organisms from becoming too warm or too cold. Explain what property of water enables it to effectively regulate an organism's temperature.

_____

_____

_____

_____

27. Water is a polar molecule. Explain why this makes water an excellent solvent.

_____

_____

_____

_____

_____

_____

28. Imagine that many people are put into a very small room and that the door is closed and locked. When the door is finally opened, what will the people do? Compare this to the way that molecules or ions move in the process of diffusion.

_____

_____

_____

_____

_____

_____

29. Explain the function of carrier molecules in the cell membrane.

_____

_____

_____

_____

_____

_____

30. It has been hypothesized that animals and plants could coexist in a sealed environment such as a glass globe. Explain what leads scientists to such a hypothesis, using equations to support your answer.

_____

_____

_____

_____

_____

_____

_____

_____

_____

_____

Answers to Unit 1 SourceBook Assessment

Multiple Choice

1. Answer: a. carbon.

2. Answer: b. 4

3. Answer: a. carbon.

4. Answer: b. water

5. Answer: a. monosaccharide.

6. Answer: c. proteins

True/False

7. Answer: b. false

Multiple Choice

8. Answer: c. proteins.

9. Answer: a. adhesion.

10. Answer: b. water.

True/False

11. Answer: b. false

12. Answer: a. true

Multiple Choice

13. Answer: c. water molecules.

14. Answer: b. active transport.

True/False

15. Answer: b. false

Multiple Choice

16. Answer: b. chloroplasts.

17. Answer: d. glucose

18. Answer: d. metabolism.

19. Answer: b. heart

20. Answer: c. regulation.

Short Response

21. Answer:

Atoms of carbon have four electrons in their outer energy level. They need a total of eight electrons to become stable. Therefore, if four different atoms each share one electron with a carbon atom, the carbon atom will have eight electrons and will be stable.

22. Answer:

An organic compound is one that contains carbon. Monomers are units of atoms within a compound; molecules made from many monomers that have been linked together are referred to as polymers. Many organic compounds are polymers.

23. Answer: Possible answers are potatoes, cereals, bread, corn, and pasta.

24. Answer:

Carbohydrates; a marathon runner would need a lot of energy while competing, and carbohydrates are the organic compounds that provide energy.

25. Answer:

DNA is located in the nucleus and contains the genetic code, which can be replicated during cell division and passed on to the next generation. RNA is also in the nucleus. However, RNA copies codes from the DNA and takes them into the cell for use in protein synthesis.

26. Answer:

Water can absorb and give off a large amount of heat without changing temperature significantly.

27. Answer:

Because it is a polar molecule, a water molecule has positive and negative ends. This causes it to attract other charged substances such as NaCl, which breaks into Na^+ and Cl^- ions and dissolves in the water.

28. Answer:

The people will move from the area where they were close together (highly concentrated) to an area where there are fewer people (lower concentration). Similarly, diffusion is the movement of molecules from an area where molecules are highly concentrated to an area where molecules are less concentrated.

29. Answer:

Carrier molecules allow active transport. Active transport is the movement of molecules into and out of a cell against a concentration gradient (from an area of low to high concentration). Carrier molecules also transport molecules that are too large to pass through the cell membrane.

30. Answer:

If you compare the equation for photosynthesis with the equation for cellular respiration, you will see that one is basically the reverse of the other. Products of photosynthesis (carbohydrates and oxygen) are the reactants of cellular respiration, and products of cellular respiration (carbon dioxide and water) are the reactants of photosynthesis. Because of this relationship between photosynthesis and cellular respiration, it has been hypothesized that plants and animals could coexist in a sealed environment such as a glass globe.

Unit 1 Extra Assessment Items

Word Usage

1. For each group of words below, write one or more sentences to show how the words are related to one another.

   a. stomata, guard cells, epidermal cells

   _____

   _____

   _____

   b. living things, energy, respiration, raw materials, products

   _____

   _____

   _____

   c. greenhouse effect, carbon dioxide, average temperature

   _____

   _____

   _____

2. Describe the process of photosynthesis, including all of the following words: *light, carbon dioxide, starch, simple sugar, oxygen, water, chlorophyll,* and *stomata.*

   _____

   _____

   _____

   _____

   _____

   _____

   _____

   _____

   _____

   _____

Correction/Completion

3. Correct or complete the statements below.

   a. Water moves into a tree from the soil as a result of osmosis.

   _____

   _____

   b. Light is the energy source used by plants to produce starch, but to absorb the light, a plant must contain a pigment called

   _____

4. Complete the following word equation for respiration to describe what happens when you eat a hamburger:

   hamburger + _____ → _____ + _____ + _____

Short Essay

5. Describe how the animals and plants in a terrarium support each other.

   _____

   _____

   _____

   _____

   _____

   _____

6. Describe how water gets from the ground to the very top of a plant.

   _____

   _____

   _____

   _____

   _____

   _____

   _____

7. Describe how human activities may be threatening the balance between oxygen and carbon dioxide in the Earth's atmosphere.

_____

_____

_____

_____

_____

_____

Short Response

8. A small bag of concentrated salt water is placed into a beaker of pure water. The bag is made of a semipermeable membrane.

   a. After an hour has passed, has the concentration of salt in the bag *increased, decreased,* or *stayed the same?*

   _____

   b. After an hour has passed, has the concentration of salt in the beaker of water *increased, decreased,* or *stayed the same?*

   _____

   c. What process has occurred?

   _____

Graphic

9. Use the graph to complete the sentence below and to answer the question that follows.

a. The amount that plant leaves photosynthesize can be determined by

measuring the amount of _____ absorbed. This graph

shows that photosynthesis is affected by three factors:

_____, _____, and

_____.

b. Does temperature affect photosynthesis *more than, less than* or *the same as* the percent of carbon dioxide?

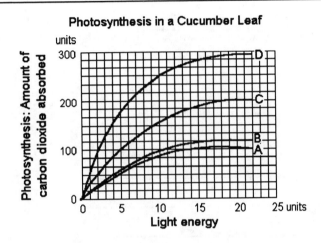

Photosynthesis in a Cucumber Leaf

Percentage of Carbon Dioxide in Atmosphere	
A 0.03% carbon dioxide at 20°C	B 0.03% carbon dioxide at 30°C
C 0.13% carbon dioxide at 20°C	D 0.13% carbon dioxide at 30°C

10. An investigation was done to study the effects of light on the production of oxygen by plants. In the first 6 hours, the plants were placed in full light. The graph shows oxygen production in the first 6 hours.

a. On the graph, show the results that you would expect if the plants were put in darkness for hours 6–12, and then returned to light for the rest of the 24-hour period.

b. On the graph, show the results that you would expect if the plants were left in full light for the whole 24-hour period.

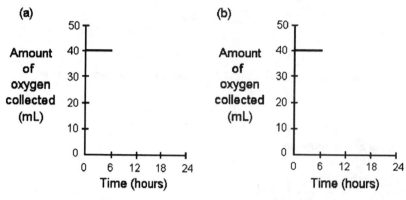

(a)

Amount of oxygen collected (mL)

(b)

Amount of oxygen collected (mL)

Illustrative

11. Use the illustration to help answer the questions that follow.

a. What is this experimental setup designed to prove?

_____

_____

_____

_____

b. Why is baking soda added to the water?

_____

_____

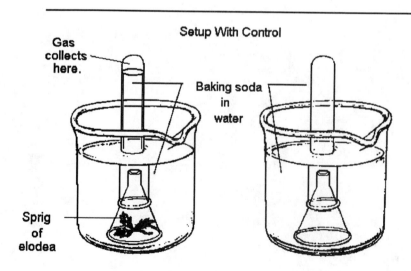

Setup With Control

Gas collects here.

Baking soda in water

Sprig of elodea

12. The drawing shows the setup for an investigation carried out over a 24-hour period. The setup is placed in bright light for 24 hours.

a. Make a new drawing showing any changes you would expect to see at the end of a 24-hour period.

b. Label the drawing to indicate the changes.

c. What process caused the change(s)?

_____

d. Which gas(es), if any, will collect in the test tube?

_____

_____

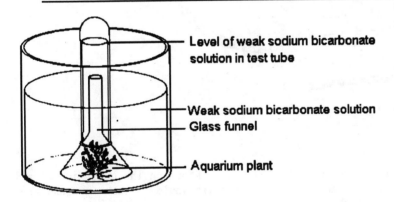

Level of weak sodium bicarbonate solution in test tube

Weak sodium bicarbonate solution

Glass funnel

Aquarium plant

13. Describe the path of energy represented by the following diagram. In your explanation, refer to each individual illustration.

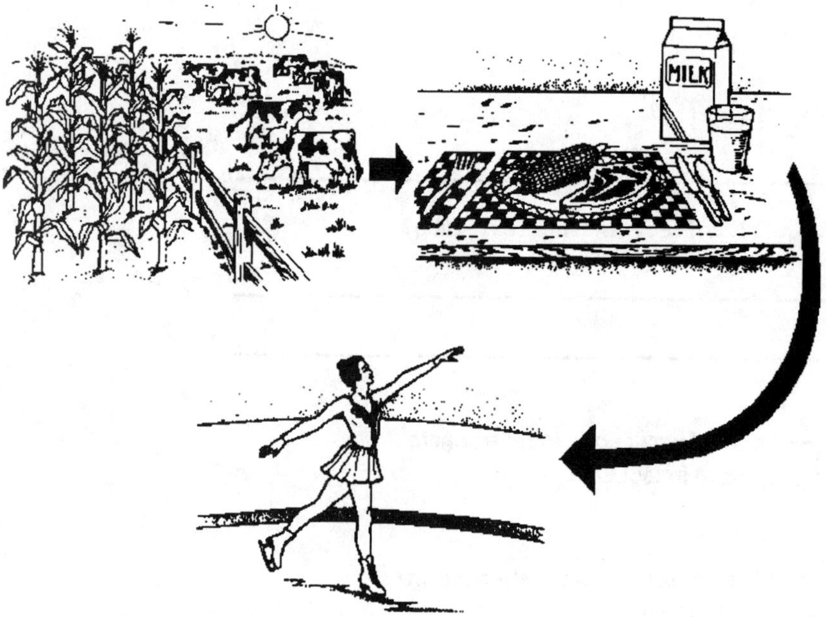

_____

_____

_____

_____

_____

_____

_____

_____

Numerical Problem

14. One page of your textbook is about 10 times thicker than a stoma on the underside of a leaf. If your textbook has 650 pages, about how many stomata would it take to equal the thickness of the pages? Show your work.

_____

_____

Performance Task

15. Complete the following activity.

Advance Preparation
Several days before you wish to perform this task, put geraniums or other suitable plants in a brightly lit location. Block the light (but not the air) from reaching half of the leaves of each plant. The night before the performance test, pick several leaves from the side of the plant that received full light, and place them in a container labeled *Y*. Then pick several leaves from the side of the plant that was blocked from the light, and place those in a container labeled *X*. Repeat for each student group. Then use a methanol bath to remove the chlorophyll from each group of leaves. **Be Careful: Methanol is poisonous and highly flammable. Keep methanol away from flames and other heat sources. Wear a lab apron and disposable gloves.**

You Will Need
* iodine solution
* an eyedropper
* leaves in a container labeled *X* and leaves in a container labeled *Y*

What to Do
The leaves in the two containers were taken from the same plant and placed in a hot methanol bath to remove the chlorophyll. Test one leaf from each container for the presence of starch. What do you observe? What might explain your observations?

Observations:

_____

_____

_____

_____

Possible explanation:

_____

_____

_____

_____

Safety Alert!

16. Complete the following activity:

You Will Need
- a microscope
- a microscope slide and coverslip
- a lettuce leaf
- a razor blade

What to Do

Prepare a wet-mount slide to observe the structure of the lettuce leaf under the microscope. Sketch what you observe. Identify the stomata and describe their function.

Sketch:

Function of the stomata:

_____

_____

_____

_____

_____

_____

_____

_____

Safety Alert!

Extended Performance Task

17. Complete the following activity:

What to Do
You are preparing to audition for a part in the television series *Hospital*. There are openings for the following parts:

Medical specialists:
- general physicians
- nurses
- pulmonary specialists (for disorders of the lungs and chest)
- cardiologists (for diseases and disorders of the heart)
- endocrinologists (for diseases and disorders of the endocrine glands and hormones)
- gastroenterologists (for diseases of the stomach and intestines)

Patients suffering from the following conditions:
- indigestion
- diabetes mellitus
- hypothyroidism
- hyperthyroidism
- coronary artery disease
- arteriosclerosis
- emphysema
- lung cancer
- asthma

Good actors research their parts well. Pair up with another person and select a condition from the second list. After doing some preliminary research, decide who will be the patient and who will be the specialist. Your audition for the part will consist of acting out an interview between the medical specialist and the patient for the purpose of diagnosing the illness.

Your preparation for the part should include
a. a diagram representing the parts of the body affected by the illness;
b. a brief description of the illness, its causes, and its relation to the body's systems;
c. a brief description of the symptoms of the illness; and
d. a brief summary of your recommendation for treating the illness.

Sources of information include encyclopedias, medical reference books, and computer network resources. You may also wish to consult a doctor or nurse.

Answers to Unit 1 Extra Assessment Items

Word Usage

1. Answer:
   a. Sample answer: The leaf surface is made up of *epidermal cells*, which form a barrier to water, and *guard cells*, which regulate the size of tiny openings called *stomata*.
   b. Sample answer: All *living things* get *energy* from the process of *respiration*, in which the *raw materials* are sugars and oxygen, and the *products* are water and carbon dioxide.
   c. Sample answer: Increasing levels of gases such as *carbon dioxide* in the atmosphere may contribute to the *greenhouse effect* and may raise the *average temperature* on Earth.

2. Answer:
   An acceptable answer should include all of the words listed. Sample answer: Photosynthesis is the process in which energy in the form of *light* enables *carbon dioxide* and *water* to combine in the presence of *chlorophyll*, producing a *simple sugar*, with *oxygen* and water vapor as byproducts. The *stomata* on the underside of the plant leaves allow carbon dioxide, oxygen, and water vapor to pass into and out of the leaves during photosynthesis. The simple sugar produced during photosynthesis can be converted and stored in the plant as *starch*. The energy stored as sugar or starch is then available to the animals that consume these plants.

Correction/Completion

3. Answer:
   a. Water moves into a tree from the soil as a result of *root pressure*.
   b. chlorophyll.

4. Answer: hamburger + <u>oxygen</u> → <u>energy</u> + <u>carbon dioxide</u> + <u>water vapor</u>

Short Essay

5. Answer:
   Sample answer: Plants obtain carbon dioxide from the air and return oxygen to it. Animals obtain oxygen from the air and return carbon dioxide to it. In other words, a mutually beneficial exchange of gases takes place between animals and plants.

6. Answer:
   Answers will vary, but should describe the diffusion of water into the plant from a region of higher concentration outside the root hair to a region of lower concentration inside the root hair. This phenomenon, called root pressure, as well as capillary action, adhesion, cohesion, and transpiration, are vital in the raising of water from the ground, through the stem, to the top of the plant, and into the air.

7. Answer:
   Sample answer: By cutting trees and otherwise reducing vegetation, humans reduce oxygen levels in the atmosphere. In addition, by burning fuels, humans increase the amount of carbon dioxide in the atmosphere. The combination of these activities threatens the balance between oxygen and carbon dioxide in the Earth's atmosphere.

Short Response

8. Answer:
   a. Decreased
   b. Increased
   c. Osmosis

9. Answer:
  a. The amount that plant leaves photosynthesize can be determined by measuring the amount of <u>carbon dioxide</u> absorbed. This graph shows that photosynthesis is affected by three factors: <u>light energy</u>, <u>temperature</u>, and <u>percentage of carbon dioxide</u>.
  b. Less than

10. Answer:

(a)

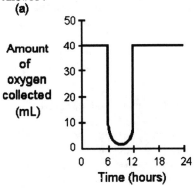

(b)

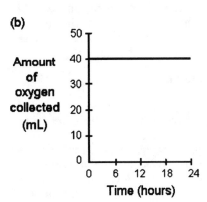

Illustrative

11. Answer:
  a. The experimental setup uses a control (the beaker without the plant) and a variable (the beaker with the plant) to show that plants release oxygen during photosynthesis.
  b. It ensures that the water contains a constant supply of carbon dioxide.

12. Answer:
  a. The drawing should show a lower level of the weak sodium bicarbonate solution in the test tube, as well as an increase in the gas collected in the top of the test tube.
  b. Students should add a new label, "Oxygen gas collected" at the top of the test tube. The label, "Level of weak sodium bicarbonate solution in test tube" will need to be moved down so that it points to the new level of the fluid in the test tube.
  c. Photosynthesis
  d. Oxygen will collect in the test tube.

13. Answer:
Sample answer: Energy from the sun is captured by the grass and the corn in the process of photosynthesis. The cows acquire some of this energy by eating the grass and then use some of that energy in their daily activities. The rest of the energy is stored. Stored energy in the corn, milk, and steak passes to the skater when she consumes these foods. She uses some of the energy from the food she has consumed as she skates.

Numerical Problem

14. Answer: 650 pages x 10 stomata per page = 6,500 stomata

Performance Task

15. Answer:
Observations:
The iodine solution turns a blue–black color when placed on the leaves from container Y, indicating the presence of starch. When placed on the leaves from container X, the iodine solution retains its reddish brown color, indicating the absence of starch.

Possible explanation:
The leaves from container Y, which received full light, could use photosynthesis to produce simple sugars, some of which were then stored as starch. The leaves from container X, which were denied full light, could not produce sugars because photosynthesis was not possible.

16. Answer:
Sketches should be clearly representative of leaf structures observed in the wet-mount slides under the microscope.

Function of the stomata:
Answers will vary. The stomata, which are located on the underside of the leaf, allow water vapor and other gases to pass into and out of the leaf during photosynthesis. Their size is regulated by the surrounding guard cells. When the guard cells absorb water, the stomata open; when water leaves the guard cells, the stomata close. The stomata usually open during the day and close at night. This schedule allows for the most successful amount of photosynthesis.

Extended Performance Task

17. Answer:
Students may choose to prepare a script that includes all of the required elements listed. They should provide evidence of their research in their diagrams and descriptions. The exchange between the medical specialist and the patient should consist of pointed questions posed by the patient and the medical specialist, such that an accurate diagnosis may be made. Recommendations should demonstrate an understanding of the physical condition as well as knowledge of the affected body systems that must be treated.

Chapter 4 Assessment

Correction/Completion

1. The statements below are incorrect or incomplete. Your challenge is to make them correct and complete.

   a. A _____ can help you develop a mental image or an idea about a complex system.

   b. When you use circumstantial evidence, you first make inferences and then make observations to support those inferences.

   _____

   _____

Short Response

2. Fill in the table below with two observations and two inferences from the following story:

   Myoshi saw a pair of cardinals in an oak tree. The first cardinal was red with black markings. The second bird wasn't as colorful. Myoshi learned that male cardinals are more colorful than females. As she watched the birds pick up sticks and leaves, she said to herself, "They must be building a nest." She thought about the baby birds that would soon be living in the nest.

Observations	Inferences

HRW material copyrighted under notice appearing earlier in this work.

49

3. "Can carrots really improve my eyesight?" Sharon asked. "Sure," her friend Patrice replied. "The Rheinhold family lives next door to me. They always eat a lot of carrots, and none of them wear glasses."

   a. What observations did Patrice make?

   _____

   _____

   _____

   _____

   b. What inference did Patrice make?

   _____

   _____

   _____

   c. Is there an error in Patrice's reasoning? Explain.

   _____

   _____

   _____

   _____

   _____

   _____

Illustrative

4. Nathan wanted to use a bicycle tire, an air pump, and two diagrams to create a model that would synthesize the following two statements: Air can be squeezed into a smaller space. If we could see air, we would see many particles. His first diagram is shown below. Next to this diagram, draw what you think his second diagram would look like.

   Flat tire

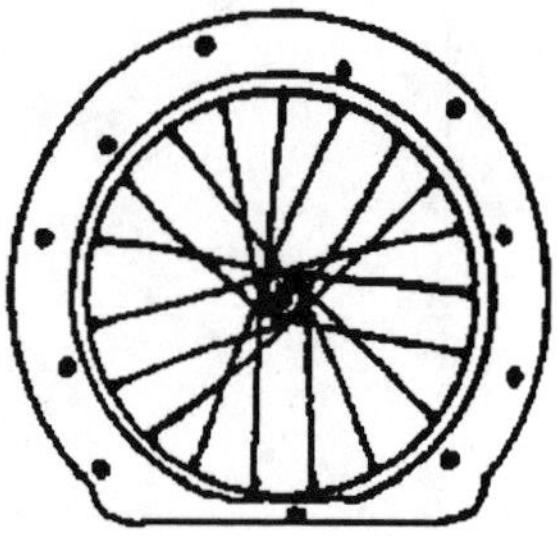

   • Particles of air

Short Essay

5. Both a flat map and a globe are used as models of the Earth. Which is the better model and why?

_____

_____

_____

_____

_____

_____

_____

_____

_____

_____

Answers to Chapter 4 Assessment

Correction/Completion

1. Answer:
   a. A <u>model</u> can help you develop a mental image or an idea about a complex system.
   b. When you use circumstantial evidence, you first make *observations* and then make *inferences* to support those *observations*.

Short Response

2. Answer:

Observations	Inferences
Sample answers: The first cardinal was red with black markings; the second wasn't as colorful.	The first bird was male; the second bird was female.
The birds picked up sticks and leaves.	Baby birds would soon live in the nest that the birds were building.

3. Answer:
   a. The Rheinhold family lives next door to her, they eat a lot of carrots, and they do not wear glasses.
   b. Patrice inferred that the reason none of the Rheinholds wear glasses is that they eat a lot of carrots.
   c. Patrice may have drawn an incorrect inference from the available observations. The fact that the Rheinhold family eats carrots may not explain why they do not wear glasses. For example, good eyesight may run in their family, or they may wear contact lenses.

Illustrative

4. Answer:

Tire after air pump is used

• Particles of air

Difficulty: 3

Short Essay

5. Answer:
Sample answer: A globe is a better model of the Earth because it is the same shape as the Earth. While both a map and a globe show where land and water exist, a globe allows you to accurately compare the placement of Earth's features. A globe also shows the tilt of the Earth's axis, and a globe can spin to simulate the Earth's rotation.

Difficulty: 4

Chapter 5 Assessment

Word Usage

1. Use the words *element*, *compound*, *atoms*, and *molecules* in one or two sentences to show how they are related.

   _____

   _____

   _____

Short Response

2. Match each event at the left with a term at the right.

   a. Dew forms when vapor in air becomes liquid.   ____ diffusion

   b. A wooden door does not close completely on a   ____ melting

      very hot day.   ____ condensation

   c. The smell of bacon frying wakes you in the   ____ evaporation

      morning.   ____ expansion

   d. A bottle of rubbing alcohol is left open; after

      a week, it is empty.

   e. A chocolate bar becomes soft on a warm day.

Data for Interpretation

3. John Dalton observed that water was made of hydrogen and oxygen. He found that 18 g of water always contained 16 g of oxygen and 2 g of hydrogen. Use Dalton's method to answer the questions about the following table:

Compound	Elements	Molecular mass	Atomic mass of elements
carbon monoxide	carbon (C) oxygen (O)	28	C = 12 O = 16
methane	carbon (C) hydrogen (H)	16	C = 12 H = 1

a. What is the ratio of carbon to oxygen in a molecule of carbon monoxide?

_____

b. There are five atoms in a molecule of methane. What is the ratio of hydrogen atoms to carbon atoms?

_____

c. What is the relationship of the mass of carbon atoms to the combined mass of hydrogen atoms in methane?

_____

_____

_____

Short Essay

4. Apply the particle model of matter to explain the following situation: In five trials, 50 g of copper is allowed to react with oxygen to form copper oxide. Each time, the copper reacts with the same amount of oxygen.

_____

_____

_____

_____

_____

_____

_____

Short Response

5. Acid rain can form when emissions of nitrogen dioxide (NO_2) and sulfur dioxide (SO_2) from factories and other sources combine with water droplets and particles of dust in the atmosphere.

a. Make an inference about elements that are used in the factories that emit these pollutants.

_____

_____

b. A model sulfur atom is made with 32 g of clay, a model nitrogen atom is made with 14 g of clay, and a model oxygen atom is made with 16 g of clay. How would the mass of sulfur compare with the mass of oxygen in a model molecule of sulfur dioxide?

_____

_____

c. Draw a model of a nitrogen dioxide molecule.

Answers to Chapter 5 Assessment

Word Usage

1. Answer:
   Sample answer: An *element* is composed of only one type of *atom*, but a *compound* is composed of *molecules* that are made of atoms of two or more elements.

Short Response

2. Answer:
   c diffusion
   e melting
   a condensation
   d evaporation
   b expansion

Data for Interpretation

3. Answer:
   a. 1 to 1
   b. 4 to 1
   c. The mass of carbon atoms is three times greater than the combined mass of the hydrogen atoms.

Short Essay

4. Answer:
   Sample answer: Each particle of copper reacts with as many particles of oxygen as it is able to, never more or less. Fifty grams of copper always contains the same number of particles. These copper particles react with a definite number of oxygen particles, and this number of oxygen particles always has the same mass.

   Difficulty: 3

Short Response

5. Answer:
   a. Students should infer that these factories use nitrogen and sulfur.
   b. At 32 g each, they would be equal.
   c. See model below.

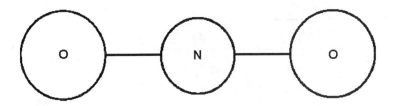

   Difficulty: 4

Chapter 6 Assessment

Word Usage

1. In one or two sentences, describe the melting of stearic acid using the following terms: *particles, change of state, faster,* and *temperature.*

   _____

   _____

   _____

   _____

   _____

Correction/Completion

2. The following sentences are incorrect or incomplete. Make the sentences correct and complete.

   a. Two different substances with the same volume always have the same mass.

   _____

   _____

   b. In general, substance that melts at 10°C has greater force of attraction among its molecules than a substance that melts at

   _____

Short Response

3. Explain why you can smell apple pie just taken out of the oven but can't smell apple pie just taken out of the refrigerator.

   _____

   _____

   _____

   _____

   _____

4. Marcel slowly heated cold water until it boiled, and he checked the temperature every minute. He recorded his results in the data table below.

Time (min.)	0	1	2	3	4	5	6
Temperature (°C)	20	25	38	76	100	100	100

a. Graph Marcel's data on the grid.

b. Why did the temperature remain the same from minutes 4 to 6?

_____

_____

_____

Short Response

5. A microwave oven works by heating water molecules inside of a substance. Use what you have learned about the particle model of matter to answer the following questions:

a. Why do potatoes sometimes explode in microwave ovens?

_____

_____

_____

b. What can a cook do to avoid having a potato explode in a microwave oven? Why would this work?

_____

_____

_____

_____

Answers to Chapter 6 Assessment

Word Usage

1. Answer:
   Sample answer: As the *temperature* rises, the *particles* of stearic acid move *faster* until the stearic acid undergoes a *change of state* from a solid to a liquid.

Correction/Completion

2. Answer:
   a. Sample answer: Two substances with the same volume *may have different masses.*
   b. any temperature lower than 10°C.

Short Response

3. Answer:
   Sample answer: Heat from the oven causes molecules from the hot pie to spread throughout the room in greater numbers and more quickly than molecules from the cold pie.

Graphic

4. Answer:
   a. See graph below.
   b. 100°C is the boiling point of water. The temperature cannot increase beyond this point.

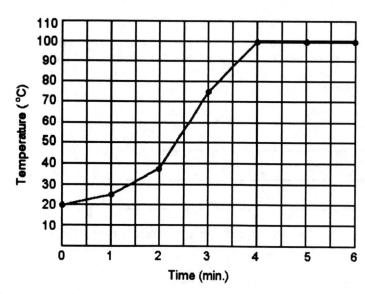

Difficulty: 3

Short Response

5. Answer:
   a. As the water molecules inside a potato are heated, they expand and become water vapor (a gas). Because the gas expands quickly, it can cause the potato to explode.
   b. By poking a few holes in the potato before heating, a cook can provide an opening for the water vapor to escape through, preventing the gas from building up inside the potato.

Difficulty: 4

Chapter 7 Assessment

Correction/Completion

1. The following sentences are incorrect or incomplete. Make the sentences correct and complete.

    a. In Rutherford's model of the atom, most of the atom's mass is contained in the electrons.

    _____

    _____

    _____

    b. An element is defined by the number of neutrons that one of its atoms has in its nucleus.

    _____

    _____

    _____

Short Response

2. How would the results of Rutherford's gold foil experiment have differed if the gold foil had been thicker?

    _____

    _____

    _____

    _____

    _____

Numerical Problem

3. Use what you've learned about atomic numbers and atomic masses to answer the following questions:

    a. An aluminum atom has an atomic number of 13 and an atomic mass of 27 amu. How many protons would an aluminum atom have? How many neutrons?

    _____

    b. Selenium has 34 protons and 45 neutrons. What is its mass in atomic mass units?

    _____

    _____

Illustrative

4. Below is one student's illustration of an atom of oxygen. Apply your knowledge of atomic models and the elements to correct the illustration in the space beside it. (The atomic number for oxygen is 8 and its atomic mass is 16 amu.)

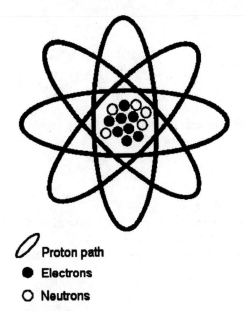

$\mathcal{O}$ Proton path

● Electrons

○ Neutrons

Short Essay

5. How does the particle theory of matter relate to how atomic models have evolved over time?

_____

_____

_____

_____

_____

_____

_____

_____

_____

_____

_____

_____

_____

Answers to Chapter 7 Assessment

Correction/Completion

1. Answer:
   a. In Rutherford's model of the atom, most of the atom's mass is contained in the *nucleus*.
   b. An element is defined by the number of *protons* in the nucleus of an atom of the element.

Short Response

2. Answer:
   Sample answer: Had the foil been thicker, there would have been more gold atoms, and therefore more nuclei and more mass, with which the alpha particles would collide and be deflected or bounced back. Fewer alpha particles would have passed through the gold foil.

Numerical Problem

3. Answer:
   a. Aluminum has 13 protons and 27 - 13, or 14, neutrons.
   b. 34 amu + 45 amu = 79 amu

Illustrative

4. Answer:

⬭ Electron path

● Protons

○ Neutrons

Difficulty: 3

Short Essay

5. Answer:
Sample answer: The scientific study of matter led to observations and inferences which suggested that all matter is composed of particles. This particle theory of matter gave way to speculation about the particles themselves. Because atoms and atomic particles are far too tiny to see, scientists created models by using objects we can actually observe or are familiar with, such as plum pudding or the solar system. Atomic models evolved as new discoveries were made by different scientists. Scientists such as Rutherford and Bohr expanded what we know about atomic particles and developed models that were built on existing models. As new discoveries continue to be made and the atomic model continues to evolve, we can further our understanding of how matter behaves according to the particle theory of matter.

Difficulty: 4

Unit 2 End-of-Unit Assessment

Word Usage

1. Explain the processes described below in terms of the particle model of matter. In each explanation, use one or more of the following words: *exothermic, endothermic, diffusion, melting, evaporation,* and *sublimation*.

   a. Butter turns to liquid when heated.

   _____

   _____

   _____

   _____

   _____

   b. Mothballs can be smelled across the room from the clothes closet in which they are located.

   _____

   _____

   _____

   _____

   c. The puddle disappears during the day.

   _____

   _____

   _____

   _____

   _____

   d. The burning candle warms the air around it.

   _____

   _____

   _____

   _____

   _____

2. Use the terms *particle model of matter*, *water*, and *inferences* in a sentence or two that explains the scientific contributions of John Dalton.

_____

_____

_____

_____

_____

_____

Correction/Completion

The following statements are either incorrect or incomplete. Your challenge is to make them correct and complete.

3. Objects of equal volumes must have equal densities.

_____

_____

_____

4. _____ shot alpha particles at gold foil to make inferences

about the structure of atoms. _____ used the resulting

atomic model as a basis for his study of the path of electrons in

atoms and created a new _____ in the process.

Short Response

5. Derrin used a liquid glue to repair a broken chair. He explained the process as follows: "The glue changes state from liquid to solid as it releases energy to the surrounding air. Once the molten glue solidifies, the process is complete." Do you agree or disagree with Derrin? Explain your reasoning.

_____

_____

_____

_____

_____

6. Make two inferences based on each of the following observations:

a. Steam is rising from a pot of water.

_____

_____

_____

b. The lemonade is cold and sweet.

_____

_____

_____

7. Name three observations you have made while studying this unit that support the particle model of matter.

_____

_____

_____

_____

_____

_____

_____

Numerical Problem

8. John Dalton realized that the mass of oxygen atoms is always 8 times the mass of hydrogen atoms in any given amount of water.

a. Suppose that a sample of water contains 12 g of hydrogen. How many grams of oxygen are in this sample?

_____

_____

b. Suppose that you react 40 g of oxygen with hydrogen to form water. How many grams of water are formed?

_____

_____

Illustrative

9. Apply the idea that all matter is made up of particles to explain an observation or inference based on the following illustration:

_____

_____

_____

_____

_____

_____

_____

_____

Short Response

10. Apply the particle model of matter to explain why the mercury in a thermometer rises when the thermometer is placed in hot water.

_____

_____

_____

_____

_____

Graphic

11. When a sample of water is cooled, its temperature will change over time in the way shown in the graph below. Place the letter that represents each statement or phrase below at the appropriate location on the graph.

   a. The freezing point of water
   b. Water has turned to ice.
   c. Liquid water is cooling down.
   d. A change of state

   Use your labeled graph and the particle model of matter to explain what is happening to the water as it cools.

   _____

   _____

   _____

   _____

   _____

   _____

   _____

   _____

   _____

   _____

   _____

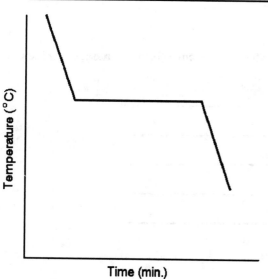

Short Essay

12. Imagine that you have been shrunk to the size of a water molecule. Write a story (one or two paragraphs) that describes an adventure you have while you are this size. Include one thing you have learned about the way particles of matter behave. Underline this idea.

_____

_____

_____

_____

_____

_____

_____

_____

_____

_____

_____

_____

_____

_____

_____

_____

Answers to Unit 2 End-of-Unit Assessment

Word Usage

1. Answer:
   Sample answers:
   a. The *melting* of butter is an *endothermic* change. When a certain amount of heat is absorbed, the particles that make up butter have enough energy to move around, so the butter melts.
   b. The particles that make up the mothballs go directly from the solid state into the gas state by a process known as *sublimation*. Once in the air, the particles spread throughout the room by the process of *diffusion*.
   c. The *evaporation* of the puddle is an *endothermic* process. Heat is absorbed by the water, causing the individual particles making up the liquid to move faster and faster until some of them gain enough energy to break loose and become gas particles.
   d. The reaction between the candle wax and oxygen is an *exothermic* change. The heat released warms the area around the candle, including the surrounding air.

2. Answer:
   Sample answer: John Dalton made *inferences* about the structure of *water* from his observations of its decomposition under an electric current. These inferences led him to develop a *particle model of matter*.

Correction/Completion

3. Answer:
   Sample answer: Objects of equal volumes *do not* have equal densities *unless they also have the same mass.*

4. Answer:
   <u>Rutherford</u> shot alpha particles at gold foil to make inferences about the structure of atoms. <u>Bohr</u> used the resulting atomic model as a basis for his study of the path of electrons in atoms and created a new <u>atomic model</u> in the process.

Short Response

5. Answer:
   Sample answer: Derrin is incorrect. The glue hardens because the water in it evaporates, leaving the solid behind. If a change of state had taken place, then a change in temperature would have had to occur and the change would have to be reversible.

6. Answer:
   a. Sample inferences: Someone is cooking dinner; the water is boiling.
   b. Sample inferences: Sugar is dissolved in the lemonade; the lemonade has been in the refrigerator.

7. Answer:
   Sample answers: Sugar dissolved in water does not significantly increase the volume of the liquid. A drop of food coloring disperses evenly in a beaker of water. In any given amount of water, the mass of oxygen atoms is always 8 times the mass of the hydrogen atoms.

Numerical Problem

8. Answer:
   a. 12 g x 8 = 96 g
   b. 40 g/8 = 5 g of hydrogen
      40 g of oxygen + 5 g of hydrogen = 45 g of water

Illustrative

9. Answer:
   Consider the answer acceptable if something seen in the illustration is explained correctly in terms of the particle model of matter. For example, differences between the three states of matter (liquid, solid, and gas) can be explained in terms of the particle model. Likewise, the particle model can be used to explain the processes of melting (of the icebergs), evaporation (of the water), or condensation (of water vapor into clouds).

   Difficulty: 3

Short Response

10. Answer:
    Sample answer: As the thermometer absorbs heat energy, the particles in the mercury move faster and faster. This causes them to move farther and farther apart. Thus, with rising temperature, the mercury takes up more space.

    Difficulty: 3

Graphic

11. Answer:
    Sample answer: The initial downward slope of the line corresponds to the time when the water is liquid. The individual particles of water are moving about freely. During this time, the water is cooling until it reaches its freezing point. The particles of water slow down more and more as the temperature decreases. The horizontal part of the line represents the freezing point of water. At this point, a change of state is occurring in which water goes from a liquid to a solid. During this time, the particles of water move closer together, slow down, and finally lose most of their range of motion. The final downward slope of the line corresponds to the time when the water is frozen and the temperature continues to decline.

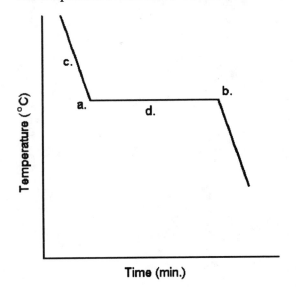

    Difficulty: 4

Short Essay

12. Answer:

Answers will vary but should demonstrate synthesis of the particle model of matter. Students may discuss changes of state as well as changes in temperature and the effects that each type of change has on particle behavior. The answer is acceptable if it describes one of the things students have learned about particle behavior in this unit.

Difficulty: 4

Unit 2 Activity Assessment

Activity Assessment

1. An Inkvestigation

Teacher's Notes
Overview
Students perform a chromatography experiment to make observations and draw inferences about the particle model of matter. They conclude by drawing a sketch to demonstrate the movement of particles in a liquid.

Materials
(per activity station)
- 3 small, wide-mouthed jars or beakers
- 3 large paper clips
- 3 pencils
- 1 white paper towel cut into 3 equal strips
- 3 markers (different colors and varieties, not waterproof)
- a metric ruler
- water
- a watch or clock

Preparation
Prior to the assessment, equip student activity stations with the materials needed.

Time Required
Each student should have 25 minutes at the activity station and an additional 15 minutes to complete his or her Data Chart.

Performance
At the end of the assessment, students should turn in the following:
- a completed Data Chart

Evaluation
The following is a recommended breakdown for evaluation of this Activity Assessment.
- 10% appropriate use of materials
- 30% ability to make observations and inferences
- 30% ability to find and apply evidence to support a theory
- 30% ability to use a sketch to clearly and accurately illustrate a scientific principle

Safety Alert!

2. An Inkvestigation

What's in ink? Since ink is a mixture, you can find out by separating some of the molecules of which it is composed. Sound impossible? Not with the simple process of chromatography! Scientists use chromatography to separate many types of mixtures. You can use it to find out what's in ink. Then show how your findings provide evidence that particles really do exist in matter.

Before You Begin . . .

As you work through the tasks, keep in mind that your teacher will be observing the following:

• how well you use the materials
• how accurately you make observations and draw inferences
• how accurately you find and apply evidence to support the particle
  model of matter
• how clearly you illustrate a principle with a sketch

Get ready to inkvestigate!

Task 1: Using a different marker for each sample, draw a straight line across each strip of paper towel (about 3 cm from the bottom). Wrap the other end of each strip around a pencil and secure it with a paper clip, as shown in the diagram. Suspend each strip in a jar filled with a small amount of water so that the line of ink is just above the water. Leave the strip in this position for 15 minutes. Watch what happens!

Task 2: Remove the strips and examine them carefully. Note the similarities and differences between samples. What do you think happened to the dyes that make up the inks? Why? Record your observations in your Data Chart on the next page.

Task 3: You have learned that all matter is made of particles and that particles come in different sizes and move at different rates. In the space provided below, use evidence from your investigation to support the particle model of matter.

Task 4: In the space provided on the next page, sketch an illustration for a child that demonstrates how the liquid particles in the inks behaved during the chromatography experiment.

Evidence for the Particle Model of Matter:

_____

_____

_____

_____

_____

_____

_____

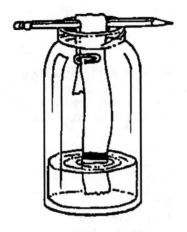

3.   Data Chart

Sample	Observations
1	
2	
3	

Sketch

Answers to Unit 2 Activity Assessment

Activity Assessment

1. Answer: Not applicable (teacher's notes)

2. Answer:
   Sample answer for Evidence for the Particle Model of Matter: Students should point out that because a single color of ink was absorbed as a blend of different colors, there is evidence that the ink was composed of different dyes. The fact that the dyes moved at different rates up the paper is evidence of the different sizes, shapes, and properties of the particles.

3. Answer:

Data Chart

Sample	Observations
1	Observations will vary depending on the type and quantity of ink used for each sample but should accurately reflect the process of paper chromatography.
2	
3	

Sketches will vary but should be logical and clearly labeled in a way that would be appropriate for a child.

Unit 2 SourceBook Assessment

Multiple Choice

1. Which of the following formulas does not represent a molecule?

    a. H_2 b. Na^- c. $C_6H_{12}O_6$ d. H_2O

2. Proton is to quark as molecule is to _____.

    a. neutron   b. compound   c. graviton   d. atom

3. Twelve is to a dozen as 6.02×10^{23} is to a _____.

    a. mole   b. gross   c. molecule   d. gallon

4. In molecules formed by ionic bonds,

    a. the ions' opposite charges attract each other.
    b. the ions' positive and negative charges repel each other.
    c. the ions bond by adhesion.
    d. the ions share electrons.

5. If you watch a thermometer in a beaker of water as the water heats, what happens to the temperature of the water as the water boils?

    a. The temperature rises.
    b. The temperature stays the same.
    c. The temperature first increases and then decreases.
    d. The temperature drops.

6. The forces between molecules in a substance are overcome when that substance is in a _____ state of matter.

    a. solid   b. liquid   c. gaseous

7. What can cause a solid to change to a liquid?

    a. an increase in kinetic energy
    b. an increase in temperature
    c. an increase in potential energy
    d. both a and b
    e. both a and c
    f. a, b, and c

8. Use the following information to figure out how many grams of carbon dioxide (CO_2) you would need to have a mole of CO_2.

Table of Atomic Mass	
H	1
C	12
O	16

a. 29 g   b. 56 g   c. 440 g   d. 44 g

Short Response

9. How many grams are in a mole of glucose ($C_6H_{12}O_6$)? (The atomic mass of H is 1; the atomic mass of C is 12; and the atomic mass of O is 16.)

_____

_____

Multiple Choice

10. Diffusion is caused by

    a. the attractive forces between particles in gases.
    b. collisions between protons and neutrons in an atom.
    c. the motions of different particles as the particles mix.
    d. gravitational forces between molecules.

True/False

11. Diffusion only occurs in gases and liquids.

    a. true   b. false

Multiple Choice

12. A waterbug can walk on the surface of a pond because of _____ between water molecules.

    a. cohesion
    b. adhesion
    c. diffusion
    d. kinetic energy

13. What force holds glue to a piece of paper?

    a. cohesion   b. fusion   c. surface tension   d. adhesion

Short Response

14. What is the basic principle behind the kinetic molecular theory of matter?

_____

_____

_____

Multiple Choice

15. In which state does water have the *least* kinetic energy?

    a. steam
    b. out of the faucet
    c. ice

16. If a container holds 20 mL of air and has a pressure of 5 atm, how can you increase the pressure to 25 atm?

    a. Reduce the volume of the container to 10 mL.
    b. Reduce the volume of the container to 4 mL.
    c. Increase the volume of the container to 15 mL.
    d. Increase the volume of the container to 100 mL.

17. Boyle's law states that the relationship between the volume of a gas and its pressure is

    a. $V_1 \times p_1 = V_2 \times p_2$.   b. $V_1 \times V_2 = p_1 \times p_2$.
    c. $V_1 / p_1 = p_2 / V_2$.   d. $V_1 \times p_2 = V_2 \times p_1$.

18. A gas can be compressed, but a liquid cannot be compressed because

    a. liquid particles have stronger surface tension than gas particles.
    b. gas particles can withstand higher pressure than liquid particles.
    c. particles in a liquid are much farther apart than particles in a gas.
    d. particles in a liquid are much closer together than particles in a gas.

Short Response

19. A hydraulic lift has an input piston that is used to apply pressure to a liquid in a closed container and an output piston that the pressure acts on. If the input piston has an area of 1 cm^2 and the output piston has an area of 5 cm^2, what force would be applied by the output piston if a force of 5 N was applied to the input piston?

    _____

    _____

    _____

    _____

Multiple Choice

20. Scientists now believe that protons and neutrons consist of three smaller particles called

    a. gravitons.   b. electrons.   c. photons.   d. quarks.

21. Which force holds the nucleus of an atom together?

    a. electromagnetic
    b. strong
    c. weak
    d. gravitational

Short Response

22. Place the following forms of matter in order from smallest to largest: *atom*, *quark*, *proton*, *molecule*, and *nucleus*.

            _____

            _____

Multiple Choice

23. Which is not true of quantum theories?

    a. They describe gravitational, strong, and weak forces.
    b. They describe electromagnetic, strong, and weak forces.
    c. They assume that only discrete packages of energy can be transferred when particles interact.
    d. None of the above

24. A single theory that describes the strong force and the electroweak force would be a(n)

    a. theory of everything.   b. grand unified theory.
    c. electroweak theory.   d. electrostrong theory.

True/False

25. The total amount of matter and energy in the universe does not change.

    a. true   b. false

Answers to Unit 2 SourceBook Assessment

Multiple Choice

1. Answer: b. Na⁻

2. Answer: d. atom

3. Answer: a. mole

4. Answer: a. the ions' opposite charges attract each other.

5. Answer: b. The temperature stays the same.

6. Answer: c. gaseous

7. Answer: d. both a and b

8. Answer: d. 44 g

Short Response

9. Answer: (6 x 12 g) + (12 x 1 g) + (6 x 16 g) = 180 g in a mole of glucose

Multiple Choice

10. Answer: c. the motions of different particles as the particles mix.

True/False

11. Answer: b. false

Multiple Choice

12. Answer: a. cohesion

13. Answer: d. adhesion

Short Response

14. Answer: Matter is made of tiny particles that are in constant motion.

Multiple Choice

15. Answer: c. ice

16. Answer: b. Reduce the volume of the container to 4 mL.

17. Answer: a. $V_1 \times p_1 = V_2 \times p_2$.

18. Answer: d. particles in a liquid are much closer together than particles in a gas.

Short Response

19. Answer:
    25 N; because the output piston is five times larger than the input piston, the force applied by the output piston is five times greater than the force applied to the input piston.

Multiple Choice

20. Answer: d. quarks.

21. Answer: b. strong

Short Response

22. Answer: Quark, proton, nucleus, atom, molecule

Multiple Choice

23. Answer: a. They describe gravitational, strong, and weak forces.

24. Answer: b. grand unified theory.

True/False

25. Answer: a. true

Unit 2 Extra Assessment Items

Word Usage

1. a. Describe the melting of salol in one or more sentences using the words *heat*, *liquid*, *particles*, and either *exothermic* or *endothermic*.

   _____

   _____

   _____

   _____

   _____

   b. In one or more sentences, describe the solidification of salol. Use the words *particles*, *solid*, *heat*, and either *exothermic* or *endothermic*.

   _____

   _____

   _____

   _____

Correction/Completion

2. Jim described the behavior of water molecules in a hot cup of tea. However, he made three mistakes. Find and correct these mistakes.

   As the water cools, the molecules move about more quickly. You can see that this must be happening because when a tea bag is dropped into the water, the brown color of the tea diffuses more slowly than it does in cold water. This shows that at a high temperature, particles move about more slowly.

   _____

   _____

   _____

   _____

   _____

   _____

Short Essay

3. Your sister, who is in the fourth grade, is studying solids, liquids, and gases. Explain to her the behavior of water molecules as an ice cube is heated to become liquid water and then steam.

_____

_____

_____

_____

_____

_____

_____

_____

_____

_____

_____

_____

Short Response

4. Use the particle model of matter to explain each of the following observations:

   a. The liquid in the thermometer rises as the temperature increases.

   _____

   _____

   _____

   _____

   b. The bathroom deodorizer disappears over time.

   _____

   _____

   _____

   _____

   _____

   c. A balloon filled with helium slowly deflates over time.

   _____

   _____

   _____

   _____

   d. Clothes hung on the clothesline dried before noon.

   _____

   _____

   _____

   _____

   e. Drink crystals dissolve faster in hot water than in cold water.

   _____

   _____

   _____

   _____

   _____

5. The following statements are part of the particle model of matter. For each one, provide an observation to support the statement.

    a. Particles making up a gas are far apart compared with those making up liquids or solids.

_____

_____

_____

    b. Adding heat to matter causes the particles to move faster.

_____

_____

_____

Graphic

6. Here is a graph showing what happens when water is heated to produce steam.

a. Which part of the graph shows the heating of liquid water with no boiling?

_____

b. Which part of the graph shows the time interval when water is boiling?

_____

c. Why does the graph curve upward past 100°C?

_____

_____

_____

_____

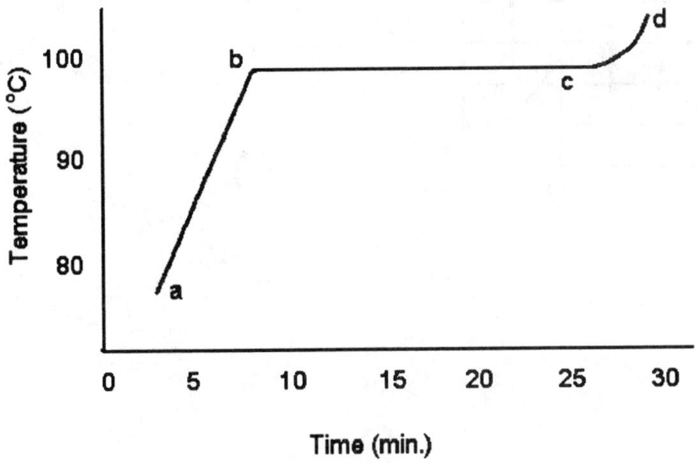

7. The following graph shows what happened as the temperature of a substance changed.

   a. Was this substance being heated or cooled?

   _____

   b. What probably happened between *b* and *c*?

   _____

   c. What probably happened between *d* and *e*?

   _____

   _____

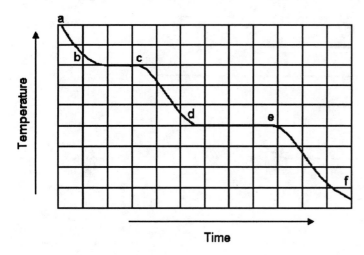

8. A Bunsen burner was placed under a beaker of ice. The Bunsen burner produced just enough heat to melt the ice and bring the water to a boil. The graph illustrates the evaporation rate (change from a liquid to a gas over time).

a. At what point does the water begin to boil? How do you know?

_____

_____

_____

_____

_____

_____

b. Draw a line on the graph showing how the evaporation rate would change if a second Bunsen burner was added after the water had already begun to boil. Explain your reasoning.

_____

_____

_____

_____

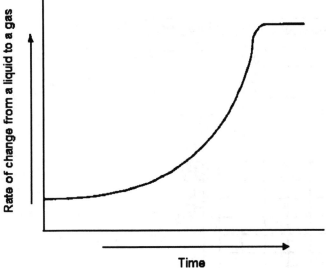

9. The graph below is an incorrect representation of the heating of naphthalene until all of it has melted. The melting point of naphthalene is 79°C. Correct the graph.

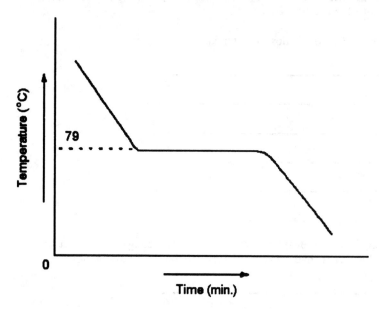

10. James collected data showing the height of a column of liquid in a test tube as the liquid was heated. His results are shown in the table below. Prepare a graph of his data. Be sure to label the vertical and horizontal axes and to title the graph. Remember to use the horizontal axis to plot the data for the variable that was manipulated.

Temperature (°C)	Height of liquid column (cm)
18	0.0
24	0.4
45	2.6
50	3.1
63	4.1

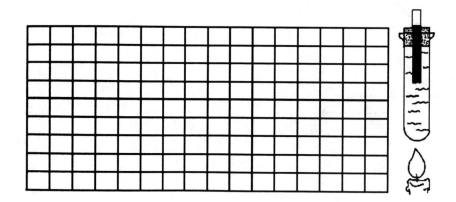

Illustrative

11. a. Use the following words to complete the concept map below: element, mixture, pure substance, matter, and compound. For the ovals containing lettered blanks, you must supply your own words.

 b. How does a pure substance differ from a mixture?

 _____

 _____

 _____

 _____

 _____

 _____

 c. How does a compound differ from an element?

 _____

 _____

 _____

 _____

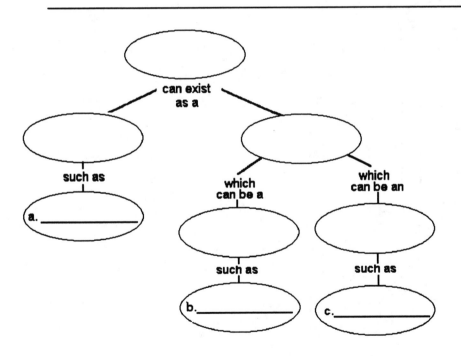

12. Exercise your scientific skills! Make three observations and two inferences based on the illustration below.

_____

_____

_____

_____

_____

_____

13. Jon completely dissolved a spoonful of sugar into a beaker containing 500 mL of water. Imagine that you could see the particles that make up the sugar and water. Draw what you think you would see.

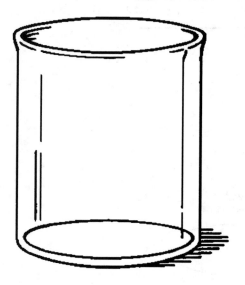

14. Imagine that you can see the particles that make up matter. Draw "before" and "after" diagrams to show what the particles in the following situations would look like:

a. You observe water molecules at 80°C and then at 120°C.

At 80°C

At 120°C

b. You observe wax molecules before and after the wax has solidified.

Melted wax

Solid wax

15. Imagine that you can see the particles that make up matter. Draw diagrams to show what the particles in the following situations would look like:
a. Forty milliliters of sugar is poured into a beaker of cold water.

b. the beaker of water in (a) 30 minutes later

16. Using modeling clay and a balance, create a model of a methane molecule. This molecule consists of four hydrogen atoms attached to a single carbon atom. The carbon atom is 12 times the mass of each hydrogen atom.

17. Observe a burning candle. Briefly discuss the following observations in terms of the particle model of matter:

    a. Wax is normally a solid.
    b. When heated, wax melts.
    c. Melted wax flows.

18. With a team of four other students, create a skit to demonstrate what is happening at the particle level in one of the following situations:

    a. melting of ice
    b. freezing of water
    c. heating of air
    d. evaporation of water
    e. expansion of iron when heated
    f. diffusion of food coloring in water

Answers to Unit 2 Extra Assessment Items

Word Usage

1. Answer:
   a. Sample answer: The melting of salol is an *endothermic* change. As the solid salol absorbs *heat*, the individual *particles* vibrate faster until they can no longer stay rigidly locked together, resulting in a *liquid*.
   b. Sample answer: The solidification of salol is an *exothermic* change. As the liquid salol gives up *heat*, the *particles* move slower and slower, until they are so close together that they form a *solid*.

Correction/Completion

2. Answer:
   Sample answer: As the water *warms*, the molecules move about more quickly. You can see that this must be happening because when a tea bag is dropped into the water, the brown color of the tea diffuses more *quickly* than it does in cold water. This shows that at a high temperature, particles move about more *quickly*.

Short Essay

3. Answer:
   Sample answer: Would you believe it? Water is actually made out of tiny particles called molecules. When the water is frozen, like in an ice cube, the molecules are stuck together. They can't move around; they can only jiggle in place. But if you heat the ice, the molecules will move back and forth faster and faster. When the ice is heated enough, some of the molecules that are moving back and forth will break loose from the solid. This is what happens when ice begins to melt. When all of the particles have broken loose from the solid, what you have is water, which is a liquid. In the water, the molecules can move around a lot more, but they still stay pretty close to each other. However, if you continue to heat the water, the molecules will start moving faster and faster. After a while, the water will begin to boil. When this happens, the molecules are moving so fast that they can move away from the other molecules in the water and escape into the air. These water molecules are known as steam.

Short Response

4. Answer:
   a. Sample answer: As the particles that make up the liquid in the thermometer begin to move faster, they move farther apart. The result is an expansion of the liquid.
   b. Sample answer: The particles that make up the deodorizer diffuse into the air.
   c. Sample answer: The helium particles slowly diffuse through the wall of the balloon.
   d. Sample answer: Because of the sun's warmth, the particles of water became water vapor and escaped from the surface of the clothes into the surrounding air.
   e. Sample answer: The hotter the water is, the faster the particles of water move. In turn, the collisions with the particles that make up the drink crystals are more frequent, allowing the crystals to dissolve faster.

5. Answer:
   a. Sample answer: You can put your hand through a liquid or a gas, but you cannot put your hand through a solid.
   b. Sample answer: When an inflated balloon is heated, the air particles inside the balloon move around faster and bump into each other more frequently. This causes the balloon to expand.

Graphic

6. Answer:
   a. The interval a→b
   b. The interval b→c
   c. The steam is being heated.

7. Answer:
   a. Cooled
   b. The substance probably changed from a gas to a liquid.
   c. The substance probably changed from a liquid to a solid.

8. Answer:
   a. The water begins to boil at the point where the graph flattens out. The liquid cannot get any hotter without turning into a gas, so you can identify the boiling point as the place where the evaporation rate has stabilized.
   b. See graph below.
   Evaporation rate increases sharply and then levels off after a second Bunsen burner is added.

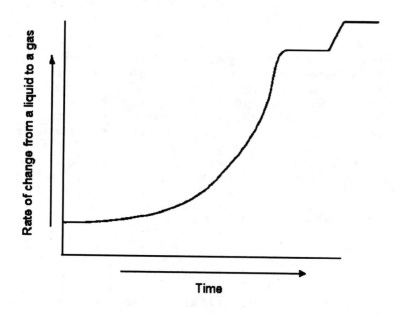

9. Answer:

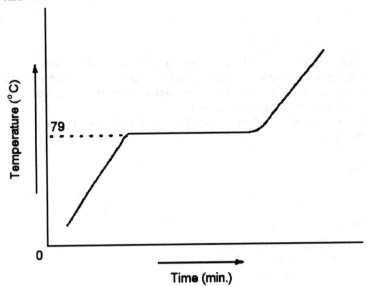

10. Answer:

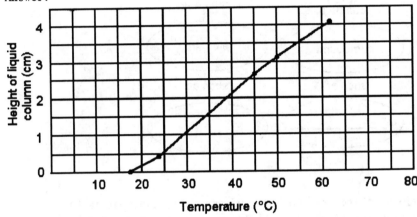

Illustrative

11. Answer:
    a. See diagram below.
    b. All pure substances on Earth can be classified as either elements or compounds. Elements cannot be broken down into components, and compounds can be separated only by chemical means. A mixture is any form of matter that contains more than one pure substance. Mixtures can be separated by physical means.
    c. A compound consists of two or more elements. An element is a substance that cannot be broken down by chemical means.

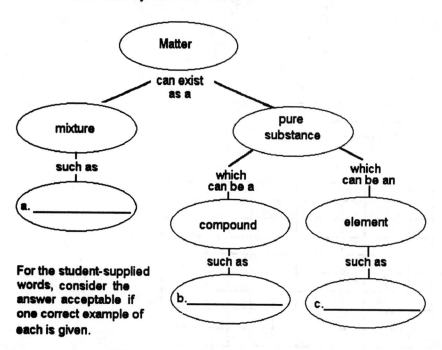

For the student-supplied words, consider the answer acceptable if one correct example of each is given.

12. Answer:
    Sample answer: Observations—there are two boys and one girl; the pool is full of water; the swimmers are standing at the edge of the pool. Inferences—the boys and girl can swim; they are at a swim meet.

13. Answer: Answers should show the sugar and water particles evenly mixed.

14. Answer:

(a)

At 80°C   At 120 °C

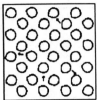   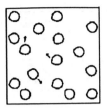

(b)   Melted wax   Solid wax

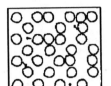   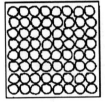

15. Answer:

(a)

(b)

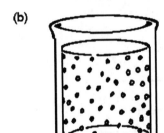

Performance Task

16. Answer:
Students should form pieces of clay in the specified ratio and secure each piece representing a hydrogen atom directly to the piece representing the carbon atom.

17. Answer:
Answers will vary. Students should show an understanding of the properties of solids and liquids according to the particle model of matter.

18. Answer:
Credit should be awarded for skits that accurately and creatively portray an understanding of the situation according to the particle model of matter.

Chapter 8 Assessment

Word Usage

1. Using the terms *machines*, *mechanical systems*, and *subsystems*, describe an automobile.

   _____

   _____

   _____

   _____

Correction/Completion

2. The following sentences are incorrect or incomplete. Your challenge is to make the sentences correct and complete.

   a. If you stand in one place for 5 minutes holding a book over your head, you are doing work because work is defined as force applied over time.

   _____

   _____

   _____

   b. The amount of energy of a moving object is affected by its size.

   _____

   _____

   _____

Short Response

3. Match each form of energy at the left with the best example on the right.

   a. Potential Energy
   b. Kinetic Energy
   c. Chemical Energy
   d. Electrical Energy

   _____ A serving of pasta provides 300 calories.

   _____ The motion of charged particles powers the fan and heating element in a hair dryer.

   _____ A diver stands at the edge of a diving tower.

   _____ Ray's pitch is moving at 120 km/h.

Numerical Problem

4. A horse can do 750 joules of work per second, giving it a power rating of 750 watts. If a car engine is rated at 175 horsepower, how many watts (joules of work per second) of power does it have? Show your work.

_____

_____

Illustrative

5. The illustration below shows the path of a bouncing ball. Match the following letters to the appropriate part of the illustration. You will use each letter more than once.

    a. potential energy
    b. kinetic energy
    c. heat energy

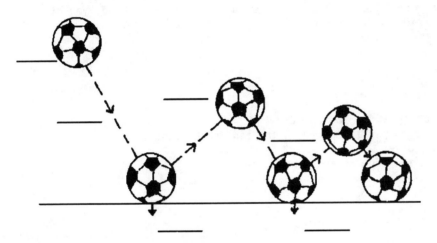

Short Response

6. Name three energy changes that affect your daily life. Be sure to include the energy converters that cause them.

_____

_____

_____

_____

_____

_____

Answers to Chapter 8 Assessment

Word Usage

1. Answer:
   Sample answer: An automobile is a complicated *mechanical system*, which is made up of *subsystems*, such as the steering system, which is made up of simpler *machines*, such as the wheels and axles.

Correction/Completion

2. Answer:
   a. If you stand in one place for 5 minutes holding a book over your head, you are *not* doing work because work is defined as force applied over *distance*.
   b. The amount of energy of a moving object is affected by its *mass and its speed*.

Short Response

3. Answer:
   c. A serving of pasta provides 300 calories.
   d. The motion of charged particles powers the fan and heating element in a hair dryer.
   a. A diver stands at the edge of a diving tower.
   b. Ray's pitch is moving at 120 km/h.

Numerical Problem

4. Answer: 750 W x 175 = 131,250 W (or 131.25 kW)

Illustrative

5. Answer:

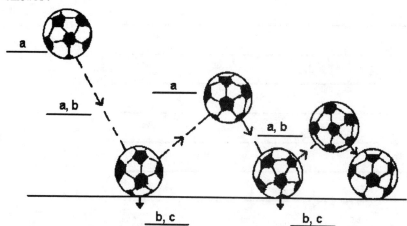

Difficulty: 3

6. Answer:
Answers will vary. Possible answers include the following: the conversion of chemical energy (food) into potential energy by the body; the change of chemical energy (gasoline) into kinetic energy by the motor of the family car; and the conversion of electrical energy into light energy by light bulbs.

Difficulty: 4

Chapter 9 Assessment

Word Usage

1. Use the words *subsystems* and *mechanical system* in one or two sentences to describe a bicycle.

   _____

   _____

2. Use the words *efficiency, work output, work input,* and *joules* in one or two sentences about a light bulb that uses 100 W of electrical energy and produces 75 W of light energy.

   _____

   _____

Correction/Completion

3. The following sentences are incorrect or incomplete. Your challenge is to make them correct and complete.

   a. The mechanical advantage of a simple machine is equal to the force exerted by the machine plus the force exerted on the machine.

   _____

   _____

   _____

   b. The smaller the force that is applied to a lever, the

   _____ the distance that the lever must move.

Short Response

4. Match each compound machine on the left with the simple machines it is composed of (listed on the right).

   a. ax   _____ 3 levers, 1 wheel and axle, 1 wedge

   b. can opener   _____ 2 levers, 1 wheel and axle

   c. scissors   _____ 2 levers, 2 wedges

   d. unicycle   _____ 1 lever, 1 wedge

Illustrative

5. Use the illustration to answer the questions about the gears.

   a. As the handle is turned clockwise, in which direction will each gear be turning?

   _____

   b. Put the gears in order from the one that will turn slowest to the one that will turn fastest.

   _____

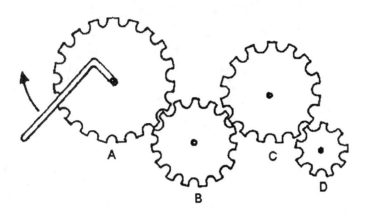

Numerical Problem

6. Lonnie turns the crank on his fishing reel to lift a fish out of the water. Assume that the fish is not struggling at all.

   a. If Lonnie is exerting a force of 2.5 N on the crank of his fishing reel, and he lifts a fish that weighs 5 N, what is the mechanical advantage of his fishing reel? Show your work.

   _____

   b. What would happen to the mechanical advantage if Lonnie doubled the force exerted?

   _____

   _____

7. Sketch a machine that could be used to crush empty metal food cans. Label your sketch to show which simple machines are included.

Explain how your simple machine works.

_____

_____

_____

_____

_____

_____

_____

Answers to Chapter 9 Assessment

Word Usage

1. Answer:
   Sample answer: A bicycle is an example of a *mechanical system*, which is composed of several *subsystems* such as a wheel and axle.

2. Answer:
   Sample answer: In 1 second, a 100 W light bulb has a *work input* of 100 J and a *work output* of 75 J, so its *efficiency* is 0.75.

Correction/Completion

3. Answer:
   a. The mechanical advantage of a simple machine is equal to the force exerted by the machine *divided by* the force exerted on the machine.
   b. longer

Short Response

4. Answer:
   __b__ 3 levers, 1 wheel and axle, 1 wedge
   __d__ 2 levers, 1 wheel and axle
   __c__ 2 levers, 2 wedges
   __a__ 1 lever, 1 wedge

Illustrative

5. Answer:
   a. *A* and *C* will turn clockwise; *B* and *D* will turn counterclockwise.
   b. *A, C, B, D*

Numerical Problem

6. Answer:
   a. Mechanical advantage = force exerted by machine + force exerted on machine = 5 N + 2.5 N = 2
   b. It would decrease by one-half (5 N + 5 N = 1).

   Difficulty: 3

Illustrative

7. Answer:
   Answers will vary but should be appropriate to the task. One approach would be a scissors-like device that has flat surfaces to crush rather than cut the cans (it would be made of two levers).

Unit 3 End-of-Unit Assessment

Word Usage

1. Describe how a wrecking ball demolishes a building using the following terms: *work*, *potential energy*, and *kinetic energy*.

   _____

   _____

   _____

   _____

2. Use the terms *kinetic energy*, *heat*, and *potential energy* to describe what happens to the energy used in pushing a heavy box up a slope.

   _____

   _____

   _____

   _____

Correction/Completion

3. Complete each statement by applying one idea you learned about work and energy.

   a. A baseball has kinetic energy when

   _____

   _____

   b. A weight lifter can measure her power by

   _____

   _____

Short Response

4. Suppose that you are able to lift only a 300 N weight. Describe how you could lift a mass of 50 kg (500 N) by using a simple machine.

_____

_____

_____

_____

_____

Graphic

5. The graph shows the relationships among the total energy, potential energy, and kinetic energy of an object in motion. Use this graph to answer the following questions.

a. What does the graph indicate about the total amount of energy an object has?

_____

_____

_____

b. What is the relationship between potential energy and kinetic energy?

_____

_____

_____

c. Identify one situation that this graph might represent.

_____

_____

_____

d. Is the graph completely accurate? Explain your answer.

_____

_____

_____

_____

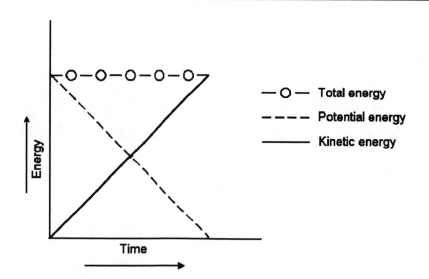

Numerical Problem

6. Use the illustration below to answer the questions that follow. Show your work.

a. What force does John exert to lift 100 kg?

_____

b. What force does Paul exert to lift 110 kg?

_____

c. Who does more work in one lift? Show your work to explain your reasoning.

_____

_____

_____

_____

Short Response

7. You and your friend Jamal are having a rope-climbing race. Your masses are both about 50 kg, and you each climb a 7 m rope. Jamal reaches the top of his rope first. He declares, "Ha! I worked harder than you, and I am the more powerful climber!" Is Jamal correct? Explain your answer.

_____

_____

_____

_____

_____

Illustrative

8. Examine the illustration below.

Will the acrobat reach the top of the 7 m pillar? Why or why not?

_____

_____

_____

_____

_____

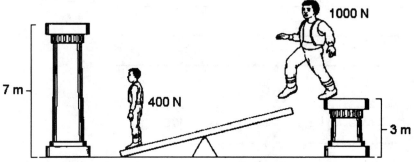

Short Response

9. Walking along the ocean shore, Amy decided to climb to the top of a nearby hill to get a view of the area.

   a. In what way was the path that Amy followed a machine?

   _____

   _____

   _____

   b. What was the advantage of climbing along a sloped path instead of climbing up a rope to the same height?

   _____

   _____

   c. It took Amy 10 minutes to reach the top of the hill. If she had taken twice as long, would this have required more or less power? Explain your answer.

   _____

   _____

   _____

Numerical Problem

10. Linda built the elevator shown in diagram A. She used a pulley, a clothesline, and a crate. However, she was unable to lift herself off the ground in this elevator (her weight plus that of the crate is 500 N).

   a. Disregarding friction, what force would she have had to exert on the rope to pull herself off the ground in diagram A?

   _____

   b. Her treehouse is 5 m above the ground. How much work would she have to do to reach the treehouse?

   _____

   Linda succeeded by adding a second pulley (diagram B).

   c. Assuming that her pulley had a mechanical advantage of 2:1, how much force did she have to exert to lift herself off the ground?

   _____

   d. How much rope did she need to pull in order to lift the elevator 5 m off the ground?

   _____

   e. How much work did she have to do to reach the tree house?

   _____

   f. Which of the two systems is more efficient? Why?

   _____

   _____

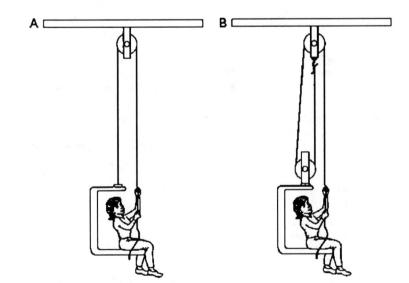

Answers to Unit 3 End-of-Unit Assessment

Word Usage

1. Answer:
   Sample answer: The *potential energy* that the ball has at the top of its swing is converted into *kinetic energy* as the wrecking ball does *work* on the building, breaking it apart. As the ball hits the building, it transfers its kinetic energy to the building.

2. Answer:
   Sample answer: The *potential energy* gained by the box as it is pushed up a slope plus the *heat* energy generated by friction must equal the *kinetic energy* of the box's movement up the slope.

Correction/Completion

3. Answer:
   a. Sample answer: . . . it flies through the air after being struck by a bat.
   b. Sample answer: . . . timing how long it takes to lift a 100 kg mass over her head.

Short Response

4. Answer:
   Sample answer: A simple machine reduces the amount of force needed to move the mass by moving that mass through a larger distance. For example, you could push, pull, or roll the mass up an inclined plane, or you could lift the mass using a pulley or a lever.

Graphic

5. Answer:
   a. The total amount of energy of an object remains the same at all times.
   b. There is an inverse relationship between potential and kinetic energy: as potential energy increases, kinetic energy decreases, and as kinetic energy increases, potential energy decreases.
   c. Sample answer: Half a pendulum swing; a spring putting a mass such as a flywheel in motion; any situation in which a mass is falling or being lowered
   d. Sample answer: The graph is not completely accurate because it does not account for heat energy. Due to friction or air resistance, some of the total energy is always converted into heat.

Numerical Problem

6. Answer:
   a. 100 kg x 10 N/kg = 1000 N
   b. 110 kg x 10 N/kg = 1100 N
   c. John does 1000 N x 2 m, or 2000 J, of work, and Paul does 1100 N x 1.75 m, or 1925 J, of work. John does more work in one lift because the product of the force he exerts and the distance through which that force is exerted is greater than that of Paul.

Short Response

7. Answer:
Jamal is only partially correct. You both did the same amount of work (50 kg x 10 N/kg x 7 m = 3500 J), because you both used the same amount of force to move the same distance. Jamal, however, did the work faster. Since power is a measurement of work done per unit of time, Jamal was the more powerful climber.

Difficulty: 3

Illustrative

8. Answer:
Yes. To reach the pillar, the lighter man needs to be given 2800 J (400 N x 7 m) of kinetic energy. When the heavier man lands on the beam, his potential energy will be converted into kinetic energy for the lighter man. The potential energy of the heavier man is 1000 N x 3 m = 3000 J, more than the 2800 J that the lighter man needs.

Difficulty: 3

Short Response

9. Answer:
   a. The hill is an inclined plane that allows Amy to decrease the force she exerts by increasing the distance through which she exerts that force.
   b. Less force is required to climb up a slope than to climb straight up.
   c. Less power, because power is the rate of doing work. Because she would have done the same amount of work in each case, only half as much power would have been required if twice as much time had been taken to climb the hill.

Difficulty: 4

Numerical Problem

10. Answer:
   a. At least 500 N
   b. 500 N x 5 m = 2500 J
   c. Half of the original force, or 500 N ÷ 2 = 250 N
   d. Twice the original distance, or 5 m x 2 = 10 m
   e. 250 N x 10 m = 2500 J
   f. The elevator in diagram A is more efficient because it has only one pulley, so it loses less energy to heat.

Difficulty: 4

Unit 3 Activity Assessment

Activity Assessment

1. **The Big Wind-up**
   Teacher's Notes
   Overview
   Students identify and diagram the simple machines they find in a wind-up toy.

   Materials
   (per activity station)
   · an inexpensive windup toy (such as an animal or car)
   · a small screwdriver

   Preparation
   Prior to the assessment, equip student activity stations with the materials needed for each experiment.

   Time Required
   Each student should have 20 minutes at the activity station and 20 minutes to complete the diagram.

   Performance
   At the end of the assessment, students should turn in the following:
   · a completed diagram that identifies potential energy, kinetic energy, and where transfer of energy occurs

   Evaluation
   The following is a recommended breakdown for evaluation of this Activity Assessment:
   · 30% ability to identify the parts of a toy and how they function
   · 30% application of knowledge of kinetic and potential energy
   · 40% ability to explain in a diagram how the mechanism of the wind-up toy transfers stored energy to make the toy move

2. The Big Wind-up

Even simple wind-up toys contain a system of machines that store and use energy to do work. At your activity station, you will find a wind-up toy. Wind it up, watch it move, and then take it apart so that you can explain how it uses simple machines in order to move.

Before you Begin . . .

As you work through the tasks, keep in mind that your teacher will be observing the following:

· how you examine the toy
· your knowledge of potential and kinetic energy
· how clearly you diagram your toy and identify the simple machines
   that make it move

Now you are ready for the big wind-up!

Task 1: Wind up the toy and watch it move on the tabletop.

Task 2: Take the toy apart and identify the internal machines that make it work. Figure out how each machine helps the toy move. In the space provided below, draw a diagram of your toy and label its internal machines.

Task 3: Identify and label the stages at which the toy uses potential and kinetic energy.

Task 4: On your diagram, show how the toy's machines transfer stored energy to make the toy move.

3.   Diagram of a Wind-up Toy

Summary of Findings

_____

_____

_____

_____

_____

Answers to Unit 3 Activity Assessment

Activity Assessment

1. Answer: Not applicable (teacher's notes)

2. Answer: Not applicable (student's notes)

3. Answer:
Diagrams will vary depending on types of wind-up toys used, but simple machines should be clearly depicted and labeled. Stages at which the toy uses potential and kinetic energy should be clearly labeled in diagrams. Student summaries should synthesize the information from the diagram and explain how the wind-up toy is able to move.

Unit 3 SourceBook Assessment

Multiple Choice

1. In a scientific sense, which of the following is NOT an example of work?

   a. jogging
   c. doing homework

   b. lifting weights
   d. climbing stairs

2. For work to be accomplished what must occur?

   a. You must get tired.
   c. You must exert a force.

   b. A job must be completed.
   d. Something must move.

3. What is the unit used to represent work?

   a. joule   b. newton   c. watt   d. meter

4. In order for movement to occur there must be

   a. a large force applied.
   c. equal forces applied.

   b. unbalanced forces.
   d. at least three forces.

5. When work is done, the forces involved are probably not constant. Therefore, when we speak of a force, we are usually talking about the

   a. average force.
   b. force applied at the halfway point.
   c. power.
   d. resistance.

True/False

6. Motion will occur as long as the effort force is equal to the resistance force.

   a. true   b. false

Multiple Choice

7. Which of the following measurements involves time?

   a. work   b. power   c. force   d. distance

8. In a 100 m race, which runner does the most work?

   a. the winner
   b. the last runner in
   c. All runners do the same amount.

9. What is the most commonly used unit for power?

   a. watt   b. milliwatt   c. kilowatt   d. deciwatt

10. The unit used to measure the amount of electricity used in a home is usually

    a. watts.   b. watt-hours.   c. joules.   d. kilowatt-hours.

11. A kilowatt-hour is a unit that measures

    a. power.   b. energy.   c. force.   d. work.

True/False

12. Simple machines are used to increase the amount of work done.

    a. true   b. false

Multiple Choice

13. A doorknob is what type of simple machine?

    a. inclined plane   b. wedge
    c. wheel and axle   d. pulley

Short Response

14. Simple machines are used to change the _____ or

   _____ of a force.

Matching

15. Match the tools on the left with the correct type of simple machine on the right.
   _____ screw   a. lever

   _____ pulley   b. inclined plane

   _____ wheelchair ramp

   _____ broom

16. Match the tool on the left with the correct class lever on the right.
   _____ scissors   a. 1st class lever

   _____ wheelbarrow   b. 2nd class lever

   _____ broom   c. 3rd class lever

   _____ baseball bat

   _____ seesaw

Short Response

17. You and a friend both push on a car that is stuck in a ditch. You push until you are both tired but the car does not move. If you consider the scientific definition of the word *work*, have you and your friend done any work? Explain your answer.

_____

_____

_____

18. If, while playing baseball, you hit the ball with a force of 100 N, and the ball travels 75 m before stopping, how much work did you do? Show your work.

_____

_____

19. Explain why large trucks have bigger steering wheels than smaller trucks or cars.

_____

_____

_____

20. If 2000 J of work is done in 25 seconds, how much power is generated? Show your work.

_____

_____

21. Use your knowledge of simple machines to solve the following problem. You are trying to load a heavy box into the back of a truck by pushing it up a ramp. After several attempts you realize the box is too heavy. If you can't find anyone to help you, how might you get the box into the truck?

_____

_____

_____

_____

22. If you used a pulley to lift a 560 N load 0.25 m off the ground, how far would you need to pull the rope if you applied a force of 35 N? Show your work.

_____

_____

23. What would be the mechanical advantage of a lever with an effort arm of 5 m and a resistance arm of 0.5 m?

_____

24. What would be the efficiency of a pulley that had a work input of 25 kJ and a work output of 20 kJ? Show your work.

_____

_____

25. What effect does friction have on the efficiency of machines?

_____

_____

_____

_____

Answers to Unit 3 SourceBook Assessment

Multiple Choice

1. Answer: c. doing homework

2. Answer: d. Something must move.

3. Answer: a. joule

4. Answer: b. unbalanced forces.

5. Answer: a. average force.

True/False

6. Answer: b. false

Multiple Choice

7. Answer: b. power

8. Answer: c. All runners do the same amount.

9. Answer: c. kilowatt

10. Answer: d. kilowatt-hours.

11. Answer: b. energy.

True/False

12. Answer: b. false

Multiple Choice

13. Answer: c. wheel and axle

Short Response

14. Answer: Simple machines are used to change the <u>size</u> or <u>direction</u> of a force.

Matching

15. Answer:
    <u>b</u> screw
    <u>a</u> pulley
    <u>b</u> wheelchair ramp
    <u>a</u> broom

16. Answer:
       a scissors
       b wheelbarrow
       c broom
       c baseball bat
       a seesaw

Short Response

17. Answer:
No. No work was done because the car did not move. Movement must occur in order for work to be done.

18. Answer: $W = F \times d$; $W = 100 \text{ N} \times 75 \text{ m}$; $W = 7500 \text{ J}$

19. Answer:
Large trucks require more force to turn. The larger steering wheel increases the distance used to turn the wheel and, thus, reduces the force required.

20. Answer: $P = W/t$; $P = 2000 \text{ J}/25 \text{ s}$; $P = 80 \text{ J/s}$, or 80 W

21. Answer:
Answers will vary. Sample answer: You could use a longer ramp. This would increase the distance the box must be moved but decrease the force needed.

22. Answer:
$F_e \times d_e = F_r \times d_r$; $35 \text{ N} \times d_e = 560 \text{ N} \times 0.25 \text{ m}$;
$d_e = (560 \text{ N} \times 0.25 \text{ m})/35 \text{ N}$; $d_e = 4.0 \text{ m}$

23. Answer: Mechanical advantage = effort arm/resistance arm; MA = 5 m/0.5 m = 10

24. Answer: Efficiency = work output/work input = 20 kJ/25 kJ = 0.8, or 80%

25. Answer: Friction decreases the efficiency of any machine.

Unit 3 Extra Assessment Items

Word Usage

1. Use the words *potential energy* and *kinetic energy* in a few sentences to describe riding a swing.

   _____

   _____

   _____

   _____

   _____

   _____

2. Use the words *work*, *power*, and *time* to compare running up a flight of stairs with walking up the stairs.

   _____

   _____

   _____

   _____

Correction/Completion

3. The following statements are either incorrect or incomplete. Your challenge is to make them correct and complete by applying what you have learned about work and energy.

   a. Mike put 35 J of work into the lifting machine that he invented to

   lift a 10 N weight a distance of 4 m. There was a loss of

   _____ J of work because _____

   _____.

   To decrease this loss, Mike would have to _____.

   b. A belt connects two wheels of different sizes. Speed is increased if the driving wheel is smaller than the driven wheel.

   _____

   _____

   _____

   c. The mechanical advantage of a unicycle can be increased by decreasing the length of the arms to the pedals.

   _____

   _____

   _____

   d. A sky diver has maximum potential energy when . . .

   _____

   _____

   _____

   e. Raj used an elevator to get to the third floor. The efficiency of

   the elevator motor can be determined by comparing _____

   _____ with _____.

4. The following statements are either incorrect or incomplete. Make them correct and complete by applying what you have learned about work and energy.

a. Bob can do work on a rock by . . .

_____

_____

b. When using an inclined plane, Sean had to apply a force of only 250 N to push a 100 kg (1000 N) box from the ground to the bed of a truck. What is the mechanical advantage of his inclined plane? Show your work.

_____

_____

Short Response

5. You have a choice between lifting a heavy box straight up to the top of a platform and pushing it there along a ramp. (Assume that friction is small.)

   a. Why might you choose to use the ramp rather than lifting the object straight up?

   _____

   _____

   _____

   b. Which of these two ways of getting the heavy box onto the platform probably requires you to use the most energy? Why?

   _____

   _____

   _____

   _____

   c. Which way requires you to exert the most force? Why?

   _____

   _____

   _____

   d. When using the ramp, how does the work input compare with the work output? Using the concept of efficiency, explain why.

   _____

   _____

   _____

   e. Why might you use a machine like a ramp to get a heavy box up to a platform rather than simply lifting the box there? Use the terms *work input*, *work output*, and *mechanical advantage* in your answer.

   _____

   _____

   _____

   _____

   _____

f. If you operate at a constant power level, would it be quicker to lift the box straight up to the platform or to push it up the ramp? Explain your answer, showing that you know the scientific meaning of the word *power*.

_____

_____

_____

_____

_____

g. Use the terms *conservation of energy, heat,* and *potential energy* to describe what happens to the energy used to slide a heavy box up a ramp.

_____

_____

_____

_____

_____

6. Hiking in the mountains, Mary decided to climb to the top of a nearby hill so that she could get a view of the area. The view from the top of the hill was magnificent. She was tired after the fast climb that took her up the straight, steep path.

a. In terms of the force required and the distance covered, how would her trip have been different on a less steep, winding path to the top?

_____

_____

_____

_____

_____

b. Suppose that Mary weighs 500 N and the hill she climbed is 200 m above the point where she began her climb. Suppose also that the path she followed is 1000 m long. How much potential energy has she acquired in climbing the hill? Show your work.

_____

_____

_____

_____

c. Would she use all of this acquired potential energy when descending the hill back to the beach? Into what forms of energy would this potential energy probably be converted as she descends the hill?

_____

_____

_____

_____

7. Will the coyote get the roadrunner? Why or why not?

_____

_____

_____

_____

_____

_____

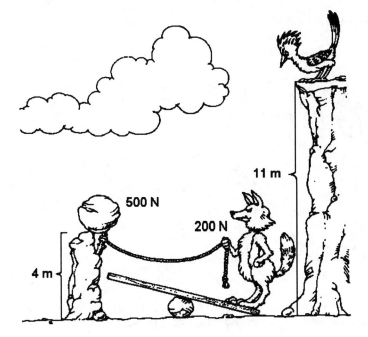

8. You wish to transfer motion by using grooved wheels or toothed gears. For each of the four items below, identify the arrangement of two parts that would accomplish the intended purpose. Identify the driving gear or wheel and the driven gear or wheel in each case.

a. to double the rate of motion

_____

b. to change the direction of motion at right angles

_____

c. to cut the rate of motion by one-half

_____

d. to increase the rate of motion three times

_____

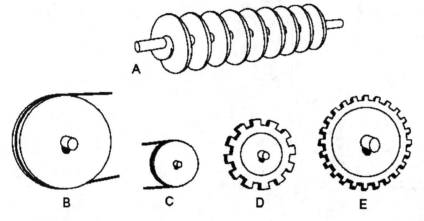

9. Sketch a machine that would enable a person in a wheelchair to get to the second floor of a building. The machine must be hand-operated—no motors are allowed.

10. Sketch a machine to help Carolyn lift a box to the second story of her house. The machine must be hand-operated—no motors are allowed.

11. Sketch a machine to help Rina and Harold move a 250 kg rock off the road. The machine must be hand-operated—no motors allowed.

12. You are building a two-story log cabin in the wilderness using only materials that you find there or that can be brought in by canoe. You must clear the land, bring the materials you need to the site (including the logs), and build the cabin. You will use a variety of tools and machines to do this, including inclined planes, levers, and pulleys. Create three sketches to show each of the following machines being used in the clearing of your campsite or in the building of the cabin:

a. an inclined plane

b. a lever

c. a pulley

Numerical Problem

13. Use the illustration below to answer the following questions.

   a. John can lift a 50 kg mass over his head from the floor 10 times in 100 seconds. What is John's power? Show your work.

   _____

   b. Paul can lift a 50 kg mass over his head from the floor 12 times in 100 seconds. What is Paul's power? Show your work.

Performance Task

14. Here is an opportunity to apply your understanding of machines, work, and energy to a practical situation. You have undoubtedly used carts, wagons, skates, skateboards, bicycles, or other forms of wheeled transport to get from one place to another. These experiences usually involve at least two machines: the wheeled vehicle you use and the slopes you climb and descend. To get started thinking about these experiences, answer the following questions:

a. Why do people usually take long, gentle slopes to get to the top of a mountain rather than taking the shortest, most direct, and steepest route? What is the advantage of taking a gentle slope to the top of a mountain? What is the disadvantage?

b. Why do people construct smooth roads and use wheeled vehicles to get from one place to another? What are some of the advantages of this method compared with walking?

Task
Using materials from the list below, find answers to the questions that follow. In each case, explain and show how you obtain your answers, including any calculations you perform.

Materials
You may carry out the tasks either outdoors, using a bathroom scale, a wheeled vehicle with a passenger, and a measuring tape, or indoors, using a spring scale, toy car or cart and load, an inclined plane (such as a board elevated at one end), and a ruler or measuring tape. In either case, you will also need a watch to measure the time in seconds.

c. What force is required to push the vehicle and its load at a steady speed up the slope? Find out if it matters whether you push it at a slow speed or at a faster speed. In which case is more power being used, at a slower or faster speed?

d. How much work is required to move the vehicle with its load from the bottom to the top of the incline? Where does the energy come from to do this work?

e. Would it take more or less force to reach the same height by lifting the vehicle and load straight up? Measure this force to find out.

f. Would it take more or less work to lift the vehicle and load straight up to the same height it reaches along the incline? Determine the amounts of work required and explain the difference, if any, in the amounts of work required along the two different paths.

g. Considering the incline as a machine, what was the efficiency of this machine?

h. Suppose that the vehicle and load roll back down the incline. Compare their potential energy and kinetic energy at the top of the incline with their potential energy and kinetic energy at the bottom.

i. What machines have you used in this task, and what advantage was gained in using them?

Answers to Unit 3 Extra Assessment Items

Word Usage

1. Answer:
   Sample answer: The swing continuously converts *kinetic energy* to *potential energy* and vice versa. When the swing reaches its highest point, kinetic energy is zero and potential energy is greatest. When the swing reaches bottom, its potential energy is zero and its kinetic energy is greatest.

2. Answer:
   Sample answer: When I walk up the stairs, I do the same amount of *work* as when I run up the stairs, but I am more tired when I run up the stairs because I do the work in less *time*. *Power* is the amount of work that you do divided by the time it takes you to do it. Therefore, running up the stairs is the more powerful activity.

Correction/Completion

3. Answer:
   a. Mike put 35 J of work into the lifting machine that he invented to lift a 10 N weight a distance of 4 m. There was a loss of <u>5</u> J of work because <u>friction turned some of the work done by the machine into heat</u>. To decrease this loss, Mike would have to <u>reduce friction</u>.
   b. Sample answer: A belt connects two wheels of different sizes. Speed is increased if the driving wheel is *larger* than the driven wheel.
   c. Sample answer: The mechanical advantage of a unicycle can be increased by *increasing* the length of the arms to the pedals.
   d. Sample answer: . . . she has just leapt from the airplane.
   e. Raj used an elevator to get to the third floor. The efficiency of the elevator motor can be determined by comparing <u>the work done on the elevator</u> with <u>the work done by the elevator</u>.

4. Answer:
   a. Sample answer: . . . lifting it into a wheelbarrow.
   b. 100 N + 250 N = 4

Short Response

5. Answer:
   a. It takes less force to push an object up a ramp than it does to lift the object straight up.
   b. Using a ramp takes more energy to get the heavy box onto the platform than does lifting it straight up because of the energy that is lost in the form of heat due to friction between the box and the ramp.
   c. It takes more force to lift the heavy box straight up because the distance through which the force is exerted is shorter.
   d. Work output is always less than work input because the ramp is not 100 percent efficient. Some energy is always lost to friction.
   e. Although the *work input* using the ramp is greater than the *work output*, there is a *mechanical advantage* to using the ramp. In other words, the force required to move the heavy box along the ramp is less than the force required to move the heavy box straight up (its weight).
   f. Sample answer: *Power* is the rate of doing work. If a person does the same amount of work every second whether moving the heavy box straight up or along the ramp, it will take longer to get the box onto the platform using the ramp. This is because using the ramp requires more work to overcome friction and, therefore, more time. Thus, it would be quicker to lift the box straight up.
   g. The law of *conservation of energy* means that the *potential energy* gained by the heavy box once it reaches the platform equals the *heat* energy generated through friction plus the kinetic energy of the heavy box as it is moved up the ramp.

6. Answer:
   a. If she had followed a less steep, winding path, the force required would have been less, but the distance covered would have been greater.
   b. The potential energy she acquired is the product of her weight and the height (not the distance) she climbed: 500 N x 200 m = 100,000 J.
   c. All of the potential energy that Mary acquired in climbing the hill would be converted primarily to kinetic energy and heat as she descends the hill.

Illustrative

7. Answer:
   The potential energy of the ball is the product of its height and weight. 4 m x 500 N = 2000 J. If all of the potential energy of the ball is converted into the kinetic energy of the coyote, the coyote will reach a height of only 10 m (2000 J ÷ 200 N = 10 m). Therefore, the coyote will not reach the roadrunner.

8. Answer:
   a. Driving gear: E; driven gear: D
   b. Driving gear: A; driven gear: D (or possibly E)
   c. Driving gear: D; driven gear: E
   d. Driving gear: C; driven gear: B

9. Answer:
   Sketches will vary. Students may draw an inclined plane or a combination of an inclined plane and a pulley system.

10. Answer: Sketches will vary. Students may draw a pulley system.

11. Answer: Sketches will vary. Students may draw a lever.

12. Answer:
   a. Sketches will vary but should be clear and logical.
   b. Sketches will vary but should be clear and logical.
   c. Sketches will vary but should be clear and logical.

Numerical Problem

13. Answer:
    a. 10 x 50 kg x 10 N/kg x 2 m ÷ 100 s = 100 W
    b. 12 x 50 kg x 10 N/kg x 1.75 m ÷ 100 s = 105 W

Performance Task

14. Answer:
    a. Answers will vary, but the students should discuss the advantages and disadvantages of exchanging force for distance and vice versa.
    b. Answers will vary, but students should discuss how friction is reduced on smooth surfaces as well as how wheeled vehicles can increase mechanical advantage.
    c. Answers will vary; the force required should be the same whether the vehicle is pushed at a slow speed or a faster speed; more power is used at a faster speed.
    d. Answers will vary. Students should be able to recognize that the energy comes from the food they eat. The chemical energy is converted to kinetic energy.
    e. It would take more force to lift the vehicle straight up.
    f. It would take less work because it covers a shorter distance.
    g. Answers will vary depending on the machines used.
    h. At the top, the potential energy would be at its greatest, and the kinetic energy would be at its lowest. At the bottom, the kinetic energy would be at its greatest, and the potential energy would be at its lowest.
    i. Sample answer for one machine: inclined plane—decreases the amount of force required

Chapter 10 Assessment

Word Usage

1. When Kurt left the airport in Miami, Florida, the temperature was 30°C, it was partly cloudy, and he was hot and sweaty. When his plane landed in Phoenix, Arizona, a few hours later, the temperature was 36°C, and the sun was shining brightly, yet he felt more comfortable than he had in Miami. Using the term *dew point*, explain why.

_____

_____

_____

_____

_____

Short Response

2. On a trip to Big Bend National Park in Texas, Portia drove from a region with a temperature of 40°C to a region with a temperature of 20°C in less than an hour. What might explain this huge change in temperature?

_____

_____

_____

_____

_____

Illustrative

3. Examine the scene below, and answer the following questions.

   a. Name the sources of atmospheric carbon dioxide in the picture.

   _____

   _____

   _____

   b. Based on the illustration, identify two ways that carbon dioxide is *removed* from the atmosphere.

   _____

   _____

   _____

   c. What could cause a change in the level of carbon dioxide in the atmosphere? What might be the consequences of such a change?

   _____

   _____

   _____

   _____

   _____

   _____

Data for Interpretation

4. Many people think distance from the sun determines a planet's temperature. However, Venus is almost twice as far from the sun as Mercury is, yet the temperature on Venus is, on average, between 50°C and 850°C warmer than the temperature on Mercury. In the space provided, use this information and the data in the table below to explain this phenomenon.

	Average distance from the sun	Average surface temperature	Gases in the atmosphere
Mercury	58 million km	450°C, day -350°C, night	none
Venus	108 million km	500°C	96% CO_2 4% N

_____

_____

_____

_____

_____

_____

_____

Graphic

5. Some scientists are not convinced that rising levels of carbon dioxide in Earth's atmosphere are causing an increase in temperature. The graph below seems to indicate that there is a connection between carbon dioxide and temperature. What might a scientist say about the data on the graph if he or she were *not* convinced that a connection between the two exists?

_____

_____

_____

_____

_____

_____

_____

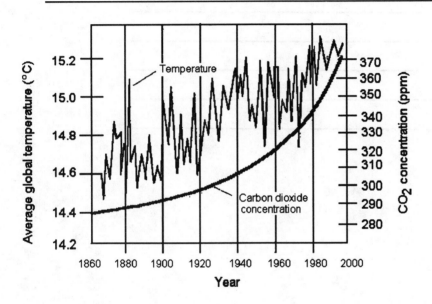

Answers to Chapter 10 Assessment

Word Usage

1. Answer:
   Sample answer: The *dew point* in Phoenix is much lower than it is in Miami, so there is less moisture in the air. As a result, perspiration evaporates more rapidly in Phoenix, transferring Kurt's body heat to his surroundings more quickly and making him feel cooler.

Short Response

2. Answer:
   Sample answer: Differences in topography and vegetation could account for the dramatic change. The park is probably characterized by large trees and steep, high mountains, while the first region is probably flat and dry. In terms of temperature, traveling 1000 m up a mountain is equivalent to traveling 1000 km north (from the equator).

Illustrative

3. Answer:
   a. The fish, the deer, the vegetation, and exhaust from the factory
   b. Carbon dioxide is removed from the atmosphere by plant life, and it is absorbed by water.
   c. Sample answer: A decrease in plant life, particularly trees, or an increase in human activity, including the burning of fossil fuels like gasoline and coal, could cause an increase in the level of carbon dioxide in the atmosphere. The consequence of the increase could be that the greenhouse effect is enhanced in the area.

Data for Interpretation

4. Answer:
   Sample answer: Mercury is closer to the sun, but Mercury has no atmosphere (and no greenhouse effect) to regulate its temperature.
   The absence of an atmosphere makes Mercury hot on the side facing the sun but cold on the opposite side because solar energy can escape into space. The atmosphere on Venus is 96 percent CO_2. This important greenhouse gas keeps the temperature on Venus at an average of 500°C even though Venus is 108 million kilometers from the sun.

   Difficulty: 3

Graphic

5. Answer:
   Sample answer: The graph shows a change of less than 1°C in about 120 years. This is not a very big change in temperature, and it is not clearly connected to the rise in carbon dioxide. For example, there is a "spike" around 1880 when the temperature was above 15°C but the carbon dioxide level was around 300 ppm. Also, there is a dramatic drop in temperature in the early 1970s when the carbon dioxide concentration was about 332 ppm.

   Difficulty: 4

Chapter 11 Assessment

Correction/Completion

1. The following sentences are incorrect or incomplete. Your challenge is to make them correct and complete.

    a. Icebergs formed from sea water tend to be more salty and more dense than the water on which they float.

    _____

    _____

    _____

    _____

    b. Ecosystems in deep-ocean canyons derive energy from the sun.

    _____

    _____

    _____

    _____

Short Response

2. Match each type of air mass with the location where it is most likely to originate.

    a. maritime polar   _____ north-central Canada

    b. maritime tropical   _____ Hudson Bay

    c. continental polar   _____ Central Africa

    d. continental tropical   _____ Caribbean Sea

3. On hot summer days, people who live near the coast often go to the beach, where ocean breezes blow from the water toward the land.

   a. What causes these breezes?

   _____

   _____

   _____

   _____

   b. What happens near the coast on a cool winter day? Explain why.

   _____

   _____

   _____

   _____

Short Essay

4. The following headline appeared in a newspaper:

   OVERLOADED SHIP SINKS AS IT SAILS INTO TROPICAL WATERS

   Use what you learned in this chapter to explain what conditions probably caused the ship to sink. Also describe how Plimsoll marks might have been used to prevent the accident.

   _____

   _____

   _____

   _____

   _____

   _____

   _____

   _____

   _____

   _____

   _____

HRW material copyrighted under notice appearing earlier in this work.

147

5. The illustration below shows a cold air mass approaching a town.

   a. Why have clouds formed at *a*?

   _____

   _____

   _____

   b. Why does the cold air mass not mix with the warm air mass?

   _____

   _____

   _____

   c. How would this situation be represented on a weather map? Illustrate your response in the space provided here.

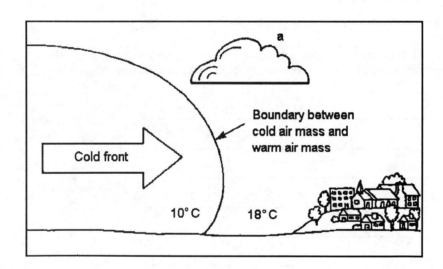

Answers to Chapter 11 Assessment

Correction/Completion

1. Answer:
   a. Sample answer: Icebergs formed from sea water tend to be *less* salty and *less* dense than the water on which they float because dissolved salt does not fit well into the crystal structure of ice.
   b. Sample answer: Ecosystems in deep-ocean canyons derive energy from *bacteria that get their energy and nutrients from the mineral-rich waters of hydrothermal vents.*

Short Response

2. Answer:
   - <u>c</u> north-central Canada
   - <u>a</u> Hudson Bay
   - <u>d</u> Central Africa
   - <u>b</u> Caribbean Sea

3. Answer:
   a. Sample answer: Warmer air over the land rises and draws in cooler air from over the water, creating a cool, onshore breeze.
   b. Sample answer: The water is warmer than the land in winter, so warmer air over the water rises and draws in cooler air from over the land creating a warm, offshore breeze.

Short Essay

4. Answer:
   Sample answer: The depth of a ship in water depends on the density of water. If a ship is loaded to its full capacity in cold water, it would sink beyond the safe limit as it entered warmer water that is less dense. Plimsoll marks provide a system for determining how full a ship can be loaded given the water conditions it will be traveling.

   Difficulty: 3

5. Answer:

   a. The warm, moist air is forced upward by the denser, cold air. As the warm air rises and cools, the moisture in it condenses, producing clouds and rain.

   b. The air masses do not mix because the cold air mass is more dense than the warm air mass.

   c. Illustrations should be similar to the following:

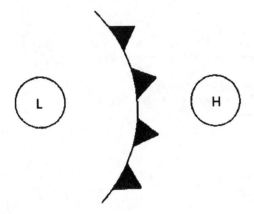

Difficulty: 4

Chapter 12 Assessment

Word Usage

1. Use the words *force* and *gravity* in an explanation of air pressure.

_____

_____

_____

Correction/Completion

2. The following sentences were offered by students during a discussion of winds and air pressure. Are they entirely correct? Improve any errors you find in each sentence.

a. In the Northern Hemisphere, winds and currents rotate in a clockwise direction because Earth's rotation, when viewed from above the North Pole, is clockwise.

_____

_____

_____

_____

b. Warm air is more dense than cool air, so air pressure at the North and South Poles is lower than air pressure at the equator.

_____

_____

_____

_____

Short Response

3. Match each item in the first column with its appropriate description in the second column.

a. barometer   _____ Very low pressure, strong winds, and heavy precipitation

b. hurricane   _____ Why winds curve clockwise or counterclockwise

c. Pascal(Pa)   _____ An instrument used to measure air pressure

d. Coriolis effect   _____ A measure of air pressure

Graphic

4. This graph shows the relationship between atmospheric pressure and altitude.

    a. State two ideas about atmospheric pressure that are suggested by the graph.

    _____

    _____

    _____

    _____

    b. If the graph were changed so that it compared latitude with atmospheric pressure, would it show the same relationship as the graph shown here? Explain.

    _____

    _____

    _____

    _____

    _____

    _____

    _____

    _____

    _____

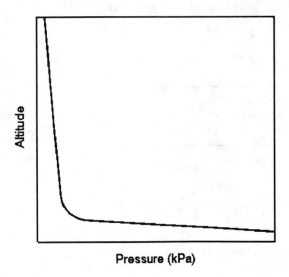

Short Essay

5. In the United States, weather systems generally move west to east. Write a paragraph explaining why this occurs. Include the concept of the Coriolis effect in your paragraph.

_____

_____

_____

_____

_____

_____

_____

_____

_____

_____

Answers to Chapter 12 Assessment

Word Usage

1. Answer:
   Sample answer: *Gravity* pulls air downward so that it exerts a *force* on the Earth's surface. This force, per unit of area, is called air pressure.

Correction/Completion

2. Answer:
   a. In the Northern Hemisphere, winds and currents rotate in a clockwise direction because, viewed from above the North Pole, Earth's rotation is *counterclockwise*.
   b. Warm air is *less* dense than cool air, so air pressure at the North and South Poles is *higher* than air pressure at the equator.

Short Response

3. Answer:
   b_ Very low pressure, strong winds, and heavy precipitation
   d_ Why winds curve clockwise or counterclockwise
   a_ An instrument used to measure air pressure
   c_ A measure of air pressure

Graphic

4. Answer:
   a. Sample answer: Atmospheric pressure is very low at high altitudes and high at low altitudes. Most of the air that makes up the atmosphere is found at lower altitudes.
   b. Sample answer: No; higher latitudes may or may not be at high altitudes, where atmospheric pressure is low, and lower latitudes may or may not be at low altitudes, where atmospheric pressure is high. Only a graph that compared latitude with temperature (temperature decreases as latitude increases) would show a relationship similar to that shown in the graph here.

   Difficulty: 3

Short Essay

5. Answer:
   Sample answer: Warm air rises at the equator and moves toward the poles. In the Northern Hemisphere, moving air or water is deflected to the right by the Coriolis effect. This causes the air moving toward the poles to move from west to east, pushing weather systems along with it.

   Difficulty: 4

Unit 4 End-of-Unit Assessment

Word Usage

1. Discuss the role of the greenhouse gases in contributing to the greenhouse effect. Use the terms *water vapor*, *atmosphere*, *carbon dioxide*, and *methane*.

   _____

   _____

   _____

   _____

2. You observe moisture forming on a cold glass of water. Use the term *dew point* in an explanation of this observation.

   _____

   _____

   _____

   _____

Correction/Completion

3. Place appropriate words in the following paragraph about ocean currents.

   In this unit you discovered two kinds of ocean currents. Deep-ocean

   currents are caused by _____ differences in ocean

   water. Surface currents are caused by _____. In the

   Northern Hemisphere, winds are deflected to the _____.

   This is due to the _____ effect.

Short Response

4. The density of ocean water varies slightly in different locations. Give a brief explanation of how each of the following processes can change the density of ocean water.

  a. evaporation

  _____

  _____

  _____

  _____

  b. freezing

  _____

  _____

  _____

  _____

5. Nina noticed that the barometric pressure was dropping, so she reminded her grandfather to bring an umbrella. Why would Nina do that?

  _____

  _____

  _____

Illustrative

6. Examine the diagram below.

   a. What phenomenon does this diagram illustrate?

   _____

   b. What is happening to the heat energy at point *A*?

   _____

   _____

   _____

   c. What is happening to the heat energy at point *B*?

   _____

   _____

   _____

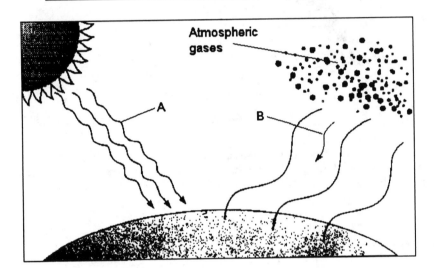

7. Here are three diagrams showing the mercury level in a barometer at locations *a*, *b*, and *c*.

   a. Which barometer diagram belongs with each location?

   _____

   _____

   b. Provide reasons for your decisions.

   _____

   _____

   _____

   _____

   _____

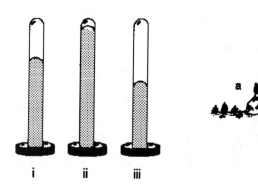

i   ii   iii

8. For *a* and *b* below, circle the toy diver that has the least pressure exerted on it. If pressure is equal, circle both divers. For *c*, the diver on the left is sinking, and the diver on the right is neither sinking nor rising. Circle the toy diver that has the greatest density.

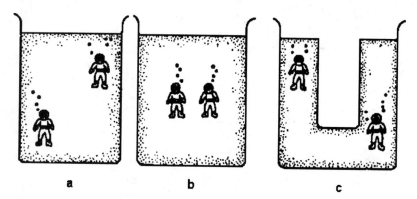

a   b   c

9. Below is a simplified diagram that represents part of the North American continent. Suppose that it is summertime, and answer the questions that follow.

a. Which location probably has the highest temperature? Explain.

_____

_____

_____

_____

b. Which location would probably have the lowest temperature? Explain.

_____

_____

_____

c. Locations *b* and *c* are at the same elevation but on opposite sides of the mountain range. Describe how this affects their temperatures.

_____

_____

_____

_____

_____

_____

Short Response

10. For each of the following situations, what happens to the dew point? Explain your reasoning.

a. A humidifier is turned on in the room.

_____

_____

_____

_____

b. A continental polar air mass replaces a maritime polar air mass.

_____

_____

_____

_____

Illustrative

11. Here is Cindy's drawing to explain why a breeze is blowing onshore. The following labels should be added to the appropriate parts of the diagram. Do this by placing the correct letters on the answer blanks provided in the diagram.

a. This heats more slowly than rock.
b. The air is less dense here.
c. Air begins to cool and become more dense.
d. The air is more dense here.

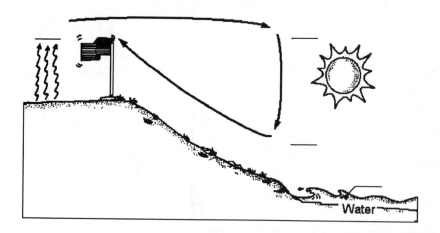

Short Essay

12. What do you think the climate might be like in the space that exists between the Earth and the moon? Explain your reasoning.

_____

_____

_____

_____

_____

_____

_____

_____

_____

_____

_____

_____

_____

_____

_____

Answers to Unit 4 End-of-Unit Assessment

Word Usage

1. Answer:
   Sample answer: The greenhouse gases—*carbon dioxide*, *methane*, and *water vapor*—are found in the Earth's *atmosphere*. They absorb and reemit heat radiating from the Earth.

2. Answer:
   Sample answer: Air that comes in contact with the cold glass is cooled. As the temperature of the air decreases and reaches the *dew point*, the moisture in the air condenses on the glass.

Correction/Completion

3. Answer:
   In this unit you discovered two kinds of ocean currents. Deep-ocean

   currents are caused by <u>density</u> differences in ocean water.

   Surface currents are caused by <u>winds</u>. In the Northern

   Hemisphere, winds are deflected to the <u>right</u>. This is due to

   the <u>Coriolis</u> effect.

Short Response

4. Answer:
   a. Evaporation can raise the density of sea water by removing some of the water and leaving the salt behind, which results in a greater density.
   b. The density of the liquid ocean water would increase if part of it froze because salt does not fit into the crystal structure of ice; salt remains in the liquid water. The frozen water would be less dense because it would contain less salt than it did as a liquid.

5. Answer:
   Sample answer: A drop in barometric pressure indicates a change in the weather—usually increasing clouds, rain, or thunderstorms.

Illustrative

6. Answer:
   a. The greenhouse effect
   b. Heat energy from the sun radiates down to Earth, where much of the heat is absorbed and re-radiated.
   c. Heat energy is absorbed by the atmospheric gases and then re-radiated back to the Earth, instead of escaping into space.

7. Answer:
   a. (i) shows the level at location a, (ii) shows the level at location c, and (iii) shows the level at location b.
   b. Sample answer: Air pressure decreases with altitude, and the lower the pressure, the shorter the column of mercury that can be supported. Therefore, the tallest column of mercury corresponds to the lowest elevation, and so on.

8. Answer:

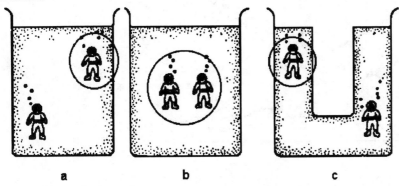

9. Answer:
   a. Location d; the moist air that blows in from the water would cool as it rises up the mountain. The drier air then warms up even more rapidly as it descends down the other side of the mountain.
   b. Location b, because it is on the side of the mountain where air from over the water is ascending
   c. Location c would probably have a warmer temperature than location b. As the air rises on the water side, the warmth from the condensing water moderates the rate at which the temperature drops. Conversely, the drier air on the opposite side of the mountain is likely to warm up more quickly as it descends.

   Difficulty: 3

Short Response

10. Answer:
    a. The dew point would decrease because more moisture is added to the air. When the air contains more moisture, that moisture condenses out of the air faster, making the dew point lower.
    b. The dew point will increase because moist air is replaced by dry air. With less moisture in the air, the dewpoint will be higher because that moisture will condense out of the air more slowly.

    Difficulty: 3

Illustrative

11. Answer:

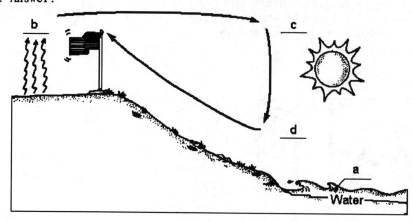

Difficulty: 4

Short Essay

12. Answer:

Answers will vary but should show an ability to synthesize the ideas of the unit. For example, Earth's climatic conditions are governed primarily by the fact that it has an atmosphere. Air pressure decreases further up into the atmosphere, as does the amount of the Earth's atmospheric gases. This affects the temperature in higher parts of the atmosphere because energy from the sun is not absorbed and reemitted by the greenhouse gases as it is on Earth. There is also no water vapor beyond Earth's atmosphere to create clouds or to influence differences in air masses. That means that there is neither rising nor falling air, nor is there wind.

Difficulty: 4

Unit 4 Activity Assessment

Activity Assessment

1. Layering Liquids
Teacher's Notes

Overview

Students determine the relative densities of several liquids by observing which ones will float on the others.

Materials
(per activity station)
- two 400 mL beakers or large jars
- a pipette
- 50 mL of corn oil
- 50 mL of glycerin
- 50 mL of water
- 50 mL of corn syrup

Preparation

Prior to the assessment, equip students activity stations with the materials needed for each experiment. Remind students not to drink any of the test liquids.

Time Required

Each student should have 20 minutes at the activity station and 15 minutes to complete the Data Sheet.

Performance

At the end of the assessment, students should turn in the following:
- a completed Data Sheet
- a jar or glass with liquids layered in it

Evaluation

The following is a recommended breakdown for evaluation of this Activity Assessment:
- 20% appropriate use of equipment and materials
- 25% use of logical experimental techniques
- 30% ability to formulate and test a prediction
- 25% ability to draw logical conclusions

2. Layering Liquids

In Unit 4 you learned about density, which determines how and why things float. In this activity you will have the chance to test your knowledge by determining the relative densities of a few common liquids.

Before You Begin . . .

As you work through the tasks, keep in mind that your teacher will be observing the following:
• how well you use the materials
• how you design and carry out the experiment
• your ability to formulate and test a prediction
• your ability to draw conclusions

Now you're ready to see what's afloat!

Task 1: Write a prediction that identifies the relative densities of the liquids at your activity station. In other words, predict which liquids will float on one another.

Task 2: Use the materials at your activity station to conduct an experiment to test your prediction. Record your observations in the Data Sheet.

Task 3: Which liquids were able to float on top of each other? How do their relative densities compare? Draw and label a sketch that illustrates the results of your experiment. Then make one suggestion about how you could improve your experiment. (You don't have to carry out the experiment again.)

3.

Predict the order in which the liquids will float, with (1) indicating the liquid that is the most dense and (4) indicating the least dense liquid.

Liquid	Predicted	Actual
water		
corn oil		
corn syrup		
glycerin		

Sketch

Improvement

_____

_____

_____

_____

Answers to Unit 4 Activity Assessment

Activity Assessment

1. Answer: Not applicable (teacher's notes)

2. Answer: Not applicable (student's notes)

3. Answer:

Data Sheet

Liquid	Predicted	Actual
water	Accept all reasonable predictions.	3
corn oil		4
corn syrup		1
glycerin		2

Improvement

Accept all reasonable answers. Improvements should demonstrate logical thought and a good understanding of the concept of density.

Sketch

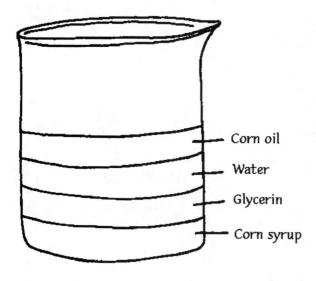

Corn oil

Water

Glycerin

Corn syrup

Unit 4 SourceBook Assessment

Multiple Choice

1. Scientists think that the first life on Earth developed over _____ years ago.

   a. 7 billion   b. 10 billion   c. 3.5 billion   d. 4.5 billion

2. Most of the oxygen present in the Earth's atmosphere is thought to have been produced from

   a. photosynthesis.   b. organic compounds.
   c. the ozone layer.   d. volcanoes.

3. What gas from Earth's early atmosphere is now largely contained in limestone deposits?

   a. oxygen   b. hydrogen   c. ammonia   d. carbon dioxide

True/False

4. There is less air pressure at high altitudes because there is not as much atmosphere above you.

   a. true   b. false

Multiple Choice

5. The layer of atmosphere where almost all of Earth's weather occurs is the

   a. stratosphere. b. troposphere. c. mesosphere. d. thermosphere.

6. The average atmospheric conditions over a long period of time is called

   a. climate.   b. wind.   c. weather.   d. precipitation.

7. What amount of the energy that reaches the Earth's atmosphere from the sun is actually absorbed by the Earth's surface?

   a. 25 percent   b. 20 percent   c. 33 percent   d. 47 percent

8. Wind moving between the equator and the poles does not move in a straight line due to the rotation of the Earth. This phenomenon is called the

   a. trade winds.   b. wind belts.
   c. Coriolis effect.   d. Bernoulli principle.

True/False

9. Warm air can hold more water vapor than can cold air.

   a. true   b. false

Multiple Choice

10. A measure of the amount of water that air contains compared with the maximum amount it could hold is referred to as

    a. humidity.
    c. saturation.
    b. relative humidity.
    d. Coriolis effect.

11. Water forming on the outside of a cold glass is called

    a. condensation. b. dew.   c. precipitation.   d. drizzle.

12. Dramatic weather changes occur because of

    a. warm fronts.   b. cold fronts.   c. Both have the same effect.

True/False

13. Scientists are not completely sure how tornadoes form.

    a. true   b. false

Multiple Choice

14. A row of thunderstorms preceding a cold front is called a

    a. cumulonimbus.
    c. source area.
    b. front.
    d. squall line.

15. Lumps containing many metals such as manganese, nickel, and even gold that are found on the ocean floor are called

    a. nodules.   b. petroleum.   c. plankton.   d. squalls.

16. Which of the following is NOT provided by the Earth's atmosphere?

    a. the gases necessary for life
    b. protection for the Earth from harmful radiation
    c. methane for photosynthesis
    d. protection from bombardment by meteors

17. After the formation of the Earth and the scattering of the original atmosphere, scientists believe that a new atmosphere was

    a. produced by the plants present on Earth.
    b. produced from gases trapped inside the rocks of the Earth's crust.
    c. transported to Earth from the sun's atmosphere.
    d. created by the ozone layer.

Matching

18. Match the descriptions on the left with the correct cloud types on the right.

_____ rain or snow producing   a. stratus

_____ puffy and white   b. cumulus

_____ flat and covering the entire sky   c. nimbostratus

_____ thin, white wisps   d. cirrus

Short Response

19. Arrange the following layers from the closest to the Earth to the farthest from the Earth: _mesosphere_, _thermosphere_, _troposphere_, and _stratosphere_.

_____

20. Explain the difference between weather and climate.

_____

_____

_____

_____

21. Explain what causes wind in our atmosphere.

_____

_____

_____

22. Explain what auroras are and what causes them.

_____

_____

_____

_____

_____

23. Why do most tornadoes occur in the spring?

_____

_____

_____

_____

_____

24. Describe the differences in the ways that rain, sleet, and hail are produced.

_____

_____

_____

_____

_____

25. Name at least one way that overharvesting of fish could affect your life.

_____

_____

_____

_____

_____

Answers to Unit 4 SourceBook Assessment

Multiple Choice

1. Answer: c. 3.5 billion

2. Answer: a. photosynthesis.

3. Answer: d. carbon dioxide

True/False

4. Answer: a. true

Multiple Choice

5. Answer: b. troposphere.

6. Answer: a. climate.

7. Answer: d. 47 percent

8. Answer: c. Coriolis effect.

True/False

9. Answer: a. true

Multiple Choice

10. Answer: b. relative humidity.

11. Answer: a. condensation.

12. Answer: b. cold fronts.

True/False

13. Answer: a. true

Multiple Choice

14. Answer: d. squall line.

15. Answer: a. nodules.

16. Answer: c. methane for photosynthesis

17. Answer: b. produced from gases trapped inside the rocks of the Earth's crust.

Matching

18. Answer:
    <u>c</u> rain or snow producing
    <u>b</u> puffy and white
    <u>a</u> flat and covering the entire sky
    <u>d</u> thin, white wisps

Short Response

19. Answer: Troposphere, stratosphere, mesosphere, thermosphere.

20. Answer:
    Weather is the condition of the atmosphere at a given time and place. Climate is the average weather conditions over many years at a particular place.

21. Answer:
    Wind is caused by the movement of air due to warm air rising and cooler air moving in to take its place.

22. Answer:
    Auroras are colorful light displays in the sky that result from the recapturing of electrons by ions in the ionosphere. Auroras are usually visible only near the poles because ions gather there due to the strong magnetic fields.

23. Answer:
    Most tornadoes occur in the spring because collisions of warm, moist air and cool, dry air—weather condition most likely to produce a tornado—are most common at that time of year.

24. Answer:
    Rain forms when cloud droplets collide and merge. Sleet is formed by the passage of rain through a cold layer of air, which causes the rain to freeze. Hail is caused by violent weather forcing raindrops high into clouds, where new layers of water collect on the raindrops and freeze.

25. Answer:
    Answers will vary. Possible answers include the following: the price of seafood could increase; some types of seafood could become less available; the economy could suffer if fishermen lost their jobs; and the environment could suffer if ocean food chains were disrupted.

Unit 4 Extra Assessment Items

Word Usage

1. Use all of the following terms in one or two sentences that explain the greenhouse effect: *respiration*, *volcanic*, *organic*, *decomposition*, and *carbon dioxide*.

_____

_____

_____

_____

2. Use the words *oceans* and *climate* in a sentence with the word *heat*.

_____

_____

Short Essay

3. Write a short newspaper article for the following headline. Include ideas associated with the terms *greenhouse* and *carbon dioxide*.

TEMPERATURES HIT RECORD HIGHS

_____

_____

_____

_____

_____

_____

_____

_____

_____

Short Response

4. "Deep ocean currents are caused by density differences in the water." Explain this statement by answering the following questions:

a. What conditions cause density differences in ocean water?

_____

_____

_____

b. Where in the oceans would you expect to find surface water sinking because of density differences?

_____

_____

_____

Graphic

5. Here is a graph of the concentration (parts per billion) of methane in the atmosphere for the last 10,000 years. Each dot represents data collected from a sample of ice buried at different depths in the glaciers that cover much of Greenland and Antarctica. (Note: graph not to scale)

a. Approximately when did the methane concentrations begin to rise noticeably?

_____

b. Suggest a reason for the rise in methane concentration.

_____

_____

_____

_____

c. What was the approximate methane concentration 10 years ago in parts per billion?

_____

d. Why should we be concerned with a rise in methane concentration?

_____

_____

_____

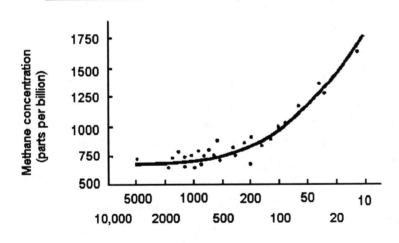

Years before present

6. Here is a graph showing atmospheric carbon dioxide concentration over the last 160,000 years.

a. Use this graph to describe the global temperature changes that might have occurred during this time. When would you expect to see that temperatures had dropped? risen?

_____

_____

_____

_____

_____

b. Give an explanation for your expectations.

_____

_____

_____

_____

_____

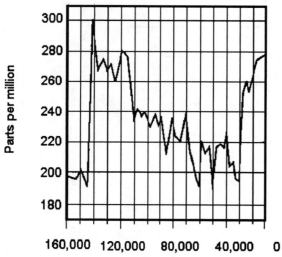

Carbon Dioxide Concentration

HRW material copyrighted under notice appearing earlier in this work.

178

7. Below is a graph of wet- and dry-bulb temperatures over a 32-hour period.

a. At what time during the 32-hour period did condensation occur?

_____

b. What was the dew point of the air at that time?

_____

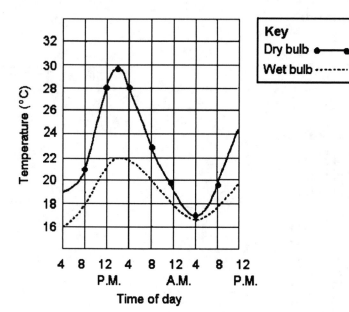

8. Blake was doing an experiment to test whether water or sand heated more quickly. He placed a heat lamp an equal distance from each substance and noted the temperatures each minute.

a. Which line represents the heating and cooling of the water? How do you know?

_____

_____

b. Which material reached the higher temperature?

_____

c. Which material cooled faster?

_____

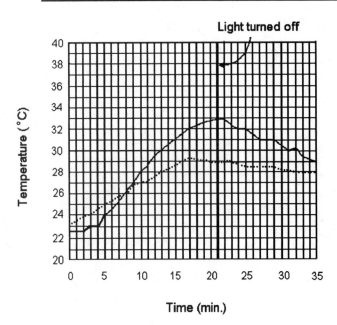

9. In 1960, the submersible *Trieste* descended to one of the deepest spots in the ocean—
the Marianas Trench, over 11,000 m below the surface.

   a. Prepare a graph of pressure versus depth using the data in this diagram.

   b. What would be an appropriate conclusion to draw from your graph?

   _____

   _____

   _____

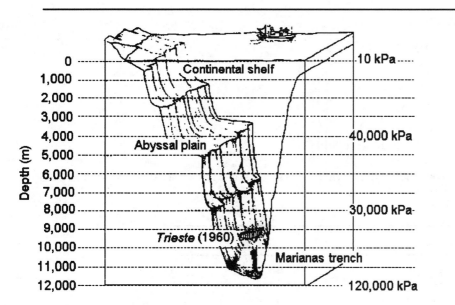

10. Dallas, Texas, has the following average temperatures:

January	10.0°C
April	20.0°C
July	33.6°C
October	23.0°C

a. Present this information in a graph.

b. Dallas lies inland. How would your graph differ for a city on the coast at the same latitude? Why?

_____

_____

_____

_____

_____

_____

Illustrative

11. Below are diagrams of a ship with the same load at two locations, one in warm, tropical waters and the other in cold, polar waters.

   a. Explain why the ship does not float at the same level at each location.

   _____

   _____

   _____

   _____

   b. Beside each of the following locations, place the letter of the appropriate diagram. Explain your answer.

   • Arctic Ocean   _____

   • Caribbean Sea   _____

   _____

   _____

   _____

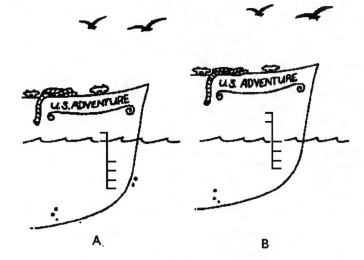

A.   B

12. After examining this sketch, answer the questions that follow.

   a. Why does the thermometer over land read a higher temperature than the one over water?

   _____

   _____

   _____

   b. Why is the air over land rising?

   _____

   _____

   c. Draw a flag on the flagpole to show the direction that you would expect the surface wind to blow.

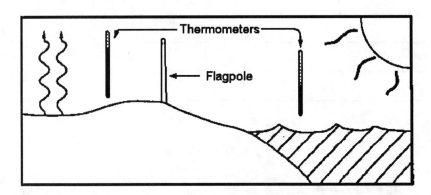

13. Here is Jane's diagram of what causes the greenhouse effect. The following labels should be added to appropriate parts of the diagram by placing each letter in the correct place on the diagram.

   a. Light from the sun is absorbed by the Earth.
   b. Longer wavelengths are reradiated from the ground.
   c. Greenhouse gases absorb and re-emit heat.

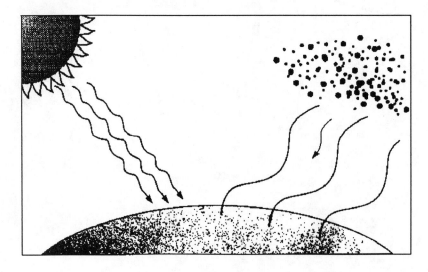

14. This is a diagram of a high-pressure area on a weather map of the Northern Hemisphere. Answer the questions that follow.

    a. What is the pressure (in kilopascals) at *A*?

    _____

    b. What is the pressure (in kilopascals) at *B*?

    _____

    c. Would the wind at *C* be blowing toward *D* or *B*? Explain.

    _____

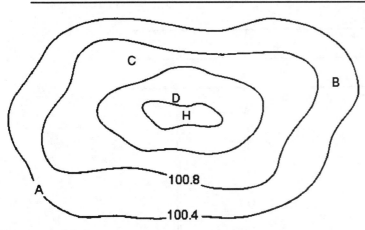

15. Jonathan drew diagrams of the paths traced by a marble on a piece of rotating poster board.

    a. Using arrows, indicate the direction that each piece of poster board is being rotated.

    b. Which one represents the Southern Hemisphere? the Northern Hemisphere? Why?

    _____

    _____

    _____

    _____

    _____

Diagram 1   Diagram 2

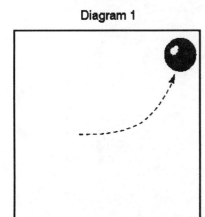
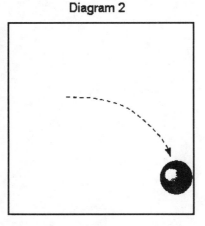

16. Water is added to containers *A*, *B*, and *C* as shown. Each container has an identical hole punched near its bottom. When the hole in container *A* is uncovered, water squirts out as shown.

   a. Complete diagrams *B* and *C* to show how far water would squirt out of each when the holes near the bottom are uncovered.

   b. Write a statement to explain what you have drawn.

   _____

   _____

   _____

   _____

   _____

   _____

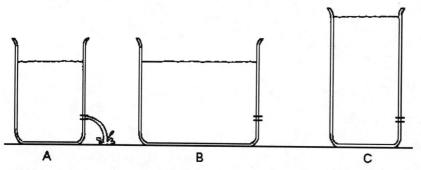

17. A small section of a weather map is drawn below. Label the following: *air mass, cold front, isobar, stationary front,* and *wind direction.*

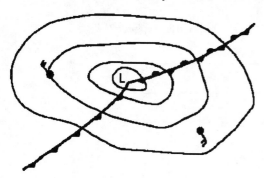

18. At the tropics there is a region of rising air called the doldrums. Air from regions north and south of the doldrums flows toward the equator. Draw arrows on the following diagram to show the direction of air moving toward the equator. Be sure to take into account the Coriolis effect.

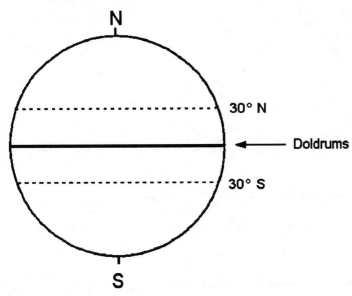

19. In the following drawings, the liquids in the flasks are separated by a piece of paper that can be removed.

Sketch what you would expect to see after the paper is removed from each setup, and give a written explanation for your prediction.

a. _____

_____

b. _____

_____

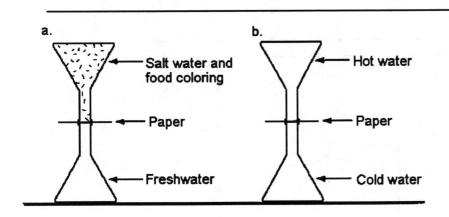

20. Draw a diagram that explains offshore breezes that occur at night. Below your diagram, explain what causes these breezes.

_____

_____

_____

_____

_____

_____

21. Holes are punched in the aluminum foil at *a* and *b*.

a. Draw arrows on the sketch to predict the direction of water flow through *a* and *b*.

b. Explain why you would expect the water to flow in this way.

_____

_____

_____

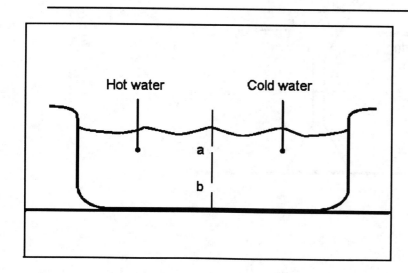

Data for Interpretation

22. You have six kinds of wood, each 50 cm^3. Use the table to answer the questions that follow.

Wood	Average density (g/cm^3)
mahogany	0.68
cedar	0.55
ebony	1.25
oak	0.82
pine	0.42
walnut	0.63

a. Which sample has the greatest mass?

_____

b. Which sample(s) would sink in water?

_____

c. Which sample would float lowest in the water?

_____

d. Which sample would float highest in the water?

_____

23. Use the density table below to answer the questions that follow.

Substance	Density (g/cm^3)
bronze	8.8
lead	11.4
tin	7.3
glass	2.6
rubber	1.2
bromine (liquid)	3.1
mercury (liquid)	13.6

a. Which substance(s) would float in mercury?

_____

b. Which substance(s) would float in bromine?

_____

c. If you had 10 cm^3 of each substance, which would have the least mass?

_____

24. Cape Hatteras, North Carolina, lies on the east coast of the United States. This table gives average water and air temperatures for each month of the year.

Month	Water temperature (°C)	Air temperature (°C)
January	13	4
February	10	4
March	10	9
April	14	15
May	18	22
June	23	27
July	25	29
August	26	29
September	25	27
October	21	20
November	16	13
December	15	7

a. Does water temperature or air temperature show the greater range?

_____

b. For a city at the same latitude, but located in the interior of the United States, how would its average temperature compare with that of Cape Hatteras? How would seasonal temperatures compare? Why?

_____

_____

_____

_____

_____

_____

c. Cape Hatteras and Santa Maria, California, lie equal distances north of the equator on opposite coasts (35°N). The average temperature of Santa Maria is 13.4°C. The average temperature of Cape Hatteras is 17.2°C. How could you explain the cooler temperatures of Santa Maria?

_____

_____

_____

_____

_____

25. Consider the data for Portland, Oregon, and Minneapolis, Minnesota. These two cities lie at approximately the same line of latitude(45°N). The table gives the average temperature for four different months.

	January	April	July	October
Portland	3.8°C	10.2°C	19.8°C	12.4°C
Minneapolis	-11.5°C	7.8°C	22.8°C	9.4°C

a. Which city has the greater range of temperatures?

_____

b. What would account for this difference?

_____

_____

_____

_____

Numerical Problem

26. Juan stands on an area that measures 0.2 m² (the surface area of his two feet). His weight is 500 N.

a. What pressure in (N/m²) is Juan exerting on the floor?

_____

b. What is the pressure in pascals?

_____

c. What is the pressure if he stands on one foot?

_____

Performance Task

27. **Materials:** eyedropper, beaker of water
Task: Examine an eyedropper. Use the scientific idea of pressure to explain how it works.

_____

_____

_____

_____

28. **Materials:** thumbtack
    Task: Examine a thumbtack. Use the scientific idea of pressure to explain how it works.

    _____

    _____

    _____

    _____

29. **Materials:** ruler, spring scale or balance, science textbook, string to wrap around the textbook
    Task: Calculate the pressure that your textbook exerts against the tabletop.

    _____

    _____

    _____

    _____

30. **Materials:** balance, graduated cylinder, salt solution
    Task: Determine the density of the salt solution.

    _____

    _____

    _____

    _____

31. **Materials:** thermometer, wet-bulb thermometer, beaker of water, dew-point table from textbook
    Task: Determine the dew point today.

    _____

    _____

    _____

    _____

32. **Materials:** straw, modeling clay, graduated cylinder, three containers of salt solutions of different concentrations identified by different food coloring
    Tasks:
    Using the straw and modeling clay, make a hydrometer that will float in water. Then arrange the three salt solutions from the greatest density to the lowest density.

Extended Performance Task

33. Using any part of a cardboard milk container, construct a boat that is stable in water. Place Plimsoll marks on the boat to indicate how much load it can take if your boat is to sail from fresh water to warm, tropical salt water.

Answers to Unit 4 Extra Assessment Items

Word Usage

1. Answer:
   Sample answer: *Carbon dioxide*, an important greenhouse gas, is added to our atmosphere as a result of *respiration* by both plants and animals, by the burning and *decomposition* of *organic* matter, and by *volcanic* eruptions.

2. Answer: Sample answer: The *oceans* influence the Earth's *climate* by storing and releasing *heat*.

Short Essay

3. Answer:
   Articles will vary but should reflect an awareness of the possible link between the concentration of carbon dioxide in the atmosphere and the intensity of the Earth's greenhouse effect.

Short Response

4. Answer:
   a. Sample answer: Differences in salt concentration and temperature cause density differences in ocean water.
   b. Sample answer: Surface water is most likely to be sinking where the air temperature is lower than that of the water.

Graphic

5. Answer:
   a. Between 200 and 500 years ago
   b. The expansion of large-scale livestock farming and other activities associated with human population expansion could be the likely cause of the increase in methane.
   c. It was approximately 1600 parts per billion 10 years ago, according to this graph.
   d. Methane is a greenhouse gas. An increase in methane in the atmosphere could cause an increase in Earth's average temperature.

6. Answer:
   a. Sample answer: You would expect temperatures to have risen sharply between 150,000 and 120,000 years ago, to have fallen gradually between 120,000 and 40,000 years ago, to rise gradually between 40,000 and 20,000 years ago, and to have risen sharply between 20,000 years ago and the present.
   b. Sample answer: The level of carbon dioxide in the atmosphere could affect the magnitude of the greenhouse effect.

7. Answer:
   a. 4 A.M.
   b. About 17°C

8. Answer:
   a. The dotted line represents the water because it heats (and cools) more slowly.
   b. Sand
   c. Sand

9. Answer:
   a. See graph below.
   b. Sample answer: Pressure increases in direct proportion to depth.

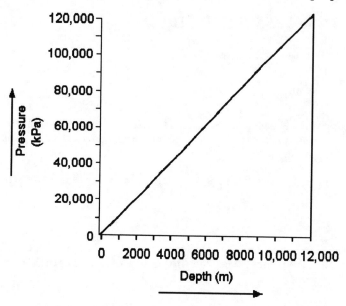

10. Answer:
    a. See graph below.
    b. Winter temperatures would be higher, while summer temperatures would be lower. The ocean absorbs and releases heat more slowly than land. This helps to moderate the temperatures near the ocean.

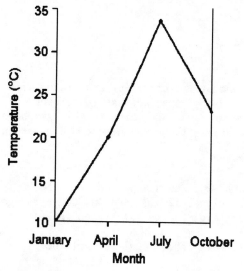

Illustrative

11. Answer:
    a. The density of water differs at each location. The denser the water, the higher the ship will float.
    b. • Arctic Ocean   B
       • Caribbean Sea   A
       The cold, polar waters of the Arctic Ocean are more dense than the warm tropical waters of the Caribbean Sea.

12. Answer:
    a. The water can absorb more of the sun's heat without an increase in temperature; the land heats up more quickly and reflects more of the sun's heat, warming the air above it.
    b. Warm air rises.
    c. Since the cooler, denser air will be blowing in from the water to replace the rising warmer air over the land, the flag should be shown blowing toward the left side of the flagpole.

13. Answer:

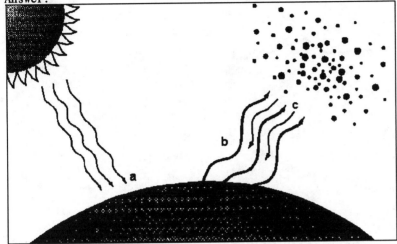

14. Answer:
    a. 100.4 kPa
    b. About 100.6 kPa
    c. The wind at *C* would be blowing toward *B* because *B* is an area of lower pressure.

15. Answer:
    a. The poster board in Diagram 1 rotates clockwise and the poster board in Diagram 2 rotates counterclockwise.
    b. Diagram 1 represents the Southern Hemisphere, where currents rotate counterclockwise. Diagram 2 represents the Northern Hemisphere, where currents rotate clockwise.

16. Answer:
    a. Diagrams should show that the water would squirt the same distance in containers *A* and *B*, but farther in container *C*.
    b. The water pressure in containers *A* and *B* should be the same; therefore the distance that the water squirts should be the same. The height of the column of water is greater in container *C*; therefore, the pressure is greater, and the distance that the water squirts out should be greater.

17. Answer:

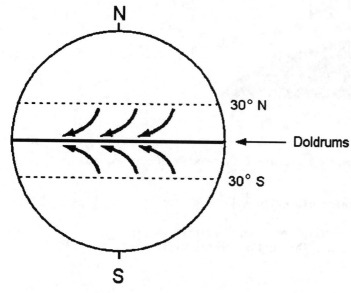

18. Answer: The diagram should look like the following:

19. Answer:
   a. Sketches should show the salt water descending and the fresh water rising. Because the salt water is denser than the fresh water, it will descend and the fresh water will rise.
   b. Sketches should show no change. Because hot water is less dense than cold water, it will remain on top.

20. Answer:
During the evening, land cools faster than water. Compared with air over the land, the air over the ocean is warmer. This warmer air over the ocean rises, and air moves in from the land to take its place.
Sample diagram:

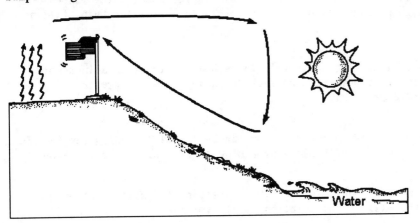

21. Answer:
a. Hot water will flow from left to right through *a*. Cold water will flow from right to left through *b*.
b. Cold water is denser than warm water so it will move through the lower hole.

Data for Interpretation

22. Answer:
a. Ebony
b. Ebony
c. Oak
d. Pine

23. Answer:
a. All of the other substances would float in mercury.
b. Rubber and glass
c. Rubber

24. Answer:
a. Air temperature
b. Sample answer: The average temperature of an inland city at the same latitude would probably be about the same as that of Cape Hatteras, but its seasonal temperatures would vary to a greater degree. Summer temperatures would be higher, and winter temperatures would be lower for the inland city. The Atlantic Ocean moderates the temperatures at Cape Hatteras.
c. Sample answer: The climate of Santa Maria could be affected by a cold offshore current, or the climate of Cape Hatteras could be affected by a warm offshore current.

25. Answer:
a. Minneapolis
b. The Pacific Ocean moderates the climate of Portland. Minneapolis, on the other hand, is far from the ocean and has a continental climate.

Numerical Problem

26. Answer:
a. 2500 N/m^2
b. 2500 Pa
c. 5000 Pa

27. Answer:
Squeezing the bulb of the eyedropper forces the air out of the tube, which creates low pressure in the tube (a partial vacuum). When the bulb is released, water rushes in from the surrounding, high pressure into the lower pressure inside the bulb, filling the partial vacuum.

28. Answer:
Pressure is force per unit area. The relatively large head of the tack allows more force to be applied to the tiny point of the tack, creating a lot of pressure so that the thumbtack can penetrate material.

29. Answer:
The pressure that the book exerts on the tabletop can be calculated using the following formula: pressure = force (weight of book) + area of book resting on the tabletop.

30. Answer:
Density is a measure of mass per unit volume. The density of the salt solution can be determined by dividing the mass of the solution by its volume.

31. Answer:
A reading should be taken with both the regular thermometer and the wet-bulb thermometer. The reading from the regular thermometer and the difference between the readings of the two thermometers can then be used to calculate the dew point using the dew-point table.

32. Answer: The hydrometer will float the highest in the solution that is the most dense.

33. Answer:
Boats will vary but should remain stable in water and should demonstrate the students' understanding of Plimsoll marks.

Chapter 13 Assessment

Word Usage

1. Write one or two sentences to explain the difference between the words *electrolyte* and *electrode*.

   _____

   _____

   _____

   _____

   _____

Correction/Completion

2. The following sentences are incorrect or incomplete. Your challenge is to make them correct and complete.

   a. When one material is rubbed against another, electricity causes charged particles to move from one material to the other.

   _____

   _____

   _____

   b. An electrostatic charge involves the continuous flow of electrons through a material such as a copper wire.

   _____

   _____

   _____

Short Response

3. Identify each of these setups with a single term.

   a. A free-floating magnetic needle used to

      tell direction   _____

   b. A group of connected cells   _____

   c. A wire coiled around a compass to

      indicate an electric current   _____

4. Circle the materials that are good conductors of electricity.

   a. glass   b. wood

   c. iron   d. zinc

   e. acid   f. plastic

Illustrative

5. Use what you've learned about static electricity to explain the illustration below.

_____

_____

_____

_____

Short Response

6. Name at least one way your life would be different if plastic were not a good insulator.

_____

_____

_____

_____

Answers to Chapter 13 Assessment

Word Usage

1. Answer:
   Sample answer: An *electrode* is the positive or negative component of a cell. It conducts electrons into or out of the *electrolyte*, a solution of chemicals that conducts electrons between the positive and negative electrodes. The conducting path causes an electric current to flow.

Correction/Completion

2. Answer:
   a. When one material is rubbed against another, *friction* causes charged particles to move from one material to the other.
   b. An *electric current* involves the continuous flow of electrons through a material such as a copper wire.

Short Response

3. Answer:
   a. A free-floating magnetic needle used to tell direction   Compass
   b. A group of connected cells   Battery
   c. A wire coiled around a compass to indicate an electric current   Galvanometer

4. Answer:
   c. iron
   d. zinc
   e. acid

Illustrative

5. Answer:
   Sample answer: The comb becomes negatively charged as it picks up electrons from the girl's hair. Her hair is now positively charged, so it is attracted to the negatively charged comb and moves toward it.

   Difficulty: 3

Short Response

6. Answer:
   Answers will vary but should reflect the fact that plastic is used to cover wiring, as on appliance cords and other items that carry electricity.

   Difficulty: 4

Chapter 14 Assessment

Correction/Completion

1. Complete the following sentence:

A(n) _____ is a unit of measure equal to 1 cycle per second.

Short Response

2. What is the difference between a dry cell and a wet cell?

_____

_____

_____

_____

3. Match each energy converter below with the description of its corresponding energy conversion.

   a. solar cell

   b. generator

   c. piezoelectric crystal

   d. thermocouple

   _____ A combination of mechanical energy and magnetism can produce very large electric currents.

   _____ Materials are stretched or squeezed to produce a current.

   _____ Heating two different metals together produces a current.

   _____ Energy from sunlight is converted into small amounts of electric current.

Illustrative

4. Name at least two ways in which this setup could be changed to produce more current.

_____

_____

_____

Short Essay

5. Imagine riding a bicycle after dark. Name one possible problem with using a bicycle dynamo instead of a battery to power the bicycle's light.

_____

_____

_____

_____

_____

_____

_____

Illustrative

6. Will this "tree battery" produce an electric current? Explain why or why not.

_____

_____

_____

_____

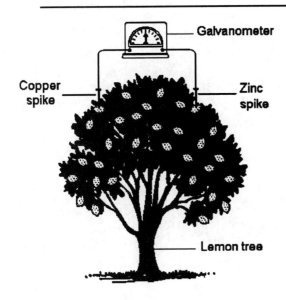

Galvanometer

Copper spike

Zinc spike

Lemon tree

Answers to Chapter 14 Assessment

Correction/Completion

1. Answer: A(n) <u>hertz</u> is a unit of measure equal to 1 cycle per second.

Short Response

2. Answer:
   In a wet cell the electrolyte is a liquid, such as salt water or sulfuric acid; in a dry cell, the electrolyte is a paste, such as ammonium chloride paste.

3. Answer:
   <u>b</u> A combination of mechanical energy and magnetism can produce very large electric currents.
   <u>c</u> Materials are stretched or squeezed to produce a current.
   <u>d</u> Heating two different metals together produces a current.
   <u>a</u> Energy from sunlight is converted into small amounts of electric current.

Illustrative

4. Answer:
   Answers will vary. Possibilities include adding more coils to the wire, moving the magnet faster, and using a more powerful magnet.

Short Essay

5. Answer:
   Answers will vary. Possible answers include the following: When the bicycle is stopped, there is no current generated, so the light goes out. The dynamo may not produce enough energy for a powerful beam.

   Difficulty: 3

Illustrative

6. Answer:
   No. The lemons that are supposed to be the electrolytes are not connected to each other, so there is no continuous path for the electrons to follow.

   Difficulty: 4

Chapter 15 Assessment

Word Usage

1. Use all of the following terms in one or two sentences to show how they are related: *amperes*, *current*, *joules*, *coulombs*, *energy*, and *volts*.

_____

_____

_____

_____

_____

Correction/Completion

2. The following sentences are incorrect or incomplete. Your challenge is to make them correct and complete.

   a. In a parallel circuit, if one appliance burns out, all of the other appliances stop working.

   _____

   _____

   b. Appliances such as toasters and irons convert _____

   energy into _____ energy because their circuits

   offer a lot of _____ .

Illustrative

3. Using the correct symbols, draw a circuit diagram to match this illustration.

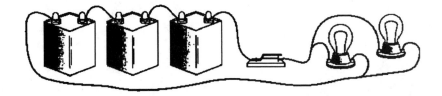

Numerical Problem

4. A 400 W hair dryer is connected to a 110 V household current.
   a. How many joules of energy are applied to each coulomb of charge?

   _____

   _____

   b. If the hair dryer is used for 3 minutes, how many joules of energy are used?

   _____

   _____

   c. How many coulombs were required during those 3 minutes?

   _____

   _____

   d. If the charge was constant during the 3 minutes, what was the rate of electrical flow per second?

   _____

   _____

Illustrative

5. Draw an illustration to match this circuit diagram.

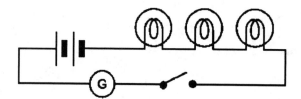

Short Essay

6. Explain why modern homes use parallel circuits rather than series circuits.

_____

_____

_____

_____

_____

_____

Answers to Chapter 15 Assessment

Word Usage

1. Answer:
   Sample answer: The rate at which electricity flows is called *current*, which is measured in *amperes* or *coulombs* per second; the *energy* given to a unit of charge flowing in a circuit is measured in *volts* or *joules* per coulomb.

Correction/Completion

2. Answer:
   a. In a *series* circuit, if one appliance burns out, all of the other appliances stop working.
   b. Appliances such as toasters and irons convert <u>electrical</u> energy into <u>heat</u> energy because their circuits offer a lot of <u>resistance</u>.

Illustrative

3. Answer: Sample answer:

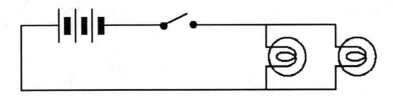

Numerical Problem

4. Answer:
   a. 110 V = 110 J/C, so 110 J of energy are applied to each coulomb of charge.
   b. 400 W = 400 J/s x 60 s/min. x 3 min. = 72,000 J of energy used in 3 min.
   c. 72,000 J ÷ 110 J/C, or about 655 coulombs, were required.
   d. 655 C/3 min. x 1 min./60 sec. = about 3.6 C/s, or 3.6 A, of current

Illustrative

5. Answer: Sample answer:

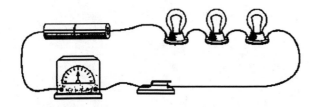

Difficulty: 3

Short Essay

6. Answer:

Sample answer: Modern homes use parallel circuits because they offer two main advantages over series circuits: current is not reduced as more appliances are added to the circuit, and the current is not broken if one appliance burns out.

Unit 5 End-of-Unit Assessment

Word Usage

1. Write a sentence about each of the situations described below. In each sentence, use at least one of these terms: *resistance, dry cells, wet cells, series, kilowatt-hours*, and *circuit*.

   a. The heating element in the toaster glows red.

   _____

   _____

   _____

   b. The meter reader checked the meter on our house today.

   _____

   _____

   _____

Correction/Completion

2. Correct the statements below.

   a. A battery consists of a single cell.

   _____

   _____

   b. Energy flow is always counterclockwise in an alternating current generator.

   _____

   _____

   c. A series circuit has at least two branches, with the current divided between each.

   _____

   _____

Short Response

3. Of the following items, circle the ones that make good conductors of electricity.

   plastic ruler   safety pin   plastic comb
   wooden pencil   scissors   nail
   plastic wrap

4. Match each of the following terms with the proper definition or description.

   a. ampere   _____ a unit of one cycle per second

   b. power   _____ a unit of measure of electrical charge

   c. hertz   _____ 1 C passing a point in 1 s

   d. coulomb   _____ the energy of the charge flowing in a circuit

   e. voltage   _____ the rate at which electrical energy is used

   f. current   _____ flow of charge in a conductor

Illustrative

5. Something is missing from each of the drawings. Sketch and label the missing parts.

a. electromagnet

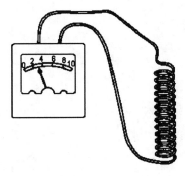

b. complete circuit

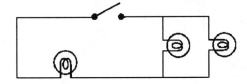

6. Use the illustrations below to answer the questions that follow:

   a. How is Circuit 1 different from Circuit 2?

   _____

   b. If you want to be able to use light A without light B, which circuit would you use?

   _____

   c. In Circuit 1, does either bulb glow more brightly than the other? If so, which one?

   _____

   d. In circuit 2, does either bulb glow more brightly than the other? If so, which one?

   _____

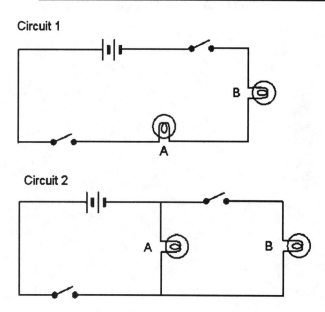

Circuit 1

Circuit 2

Numerical Problem

7. A toaster, electric kettle, and coffee maker are all being operated from the same electrical outlet.

   Information: kettle   1500 W
                     toaster   800 W
                     coffee maker   450 W
                     The house operates on a 110 V circuit.

   Would a 20 A fuse blow in this case? Show your work.

   _____

   _____

Graphic

8. Use the graph below to answer the questions that follow:

   a. Which cell has the shortest life?

   _____

   b. How many hours will the mercury cell last?

   _____

   c. Which is the strongest cell when new?

   _____

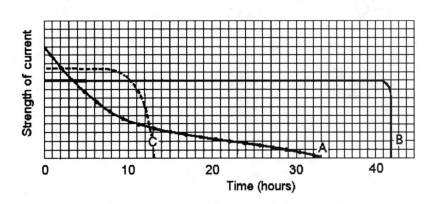

A= alkaline cell   B= mercury cell   C= ordinary cell

Illustrative

9. Using the principles of an electromagnet, design and draw a model crane that you would use to pick up thumbtacks from the floor. Clearly label your illustration to show how it works.

   Equipment: cells (2) or battery   nail
                 wires   crank
                 pulley and cable   switch

10. If necessary, correct each of the following circuits by adding, removing, or changing a wire to make a bulb light. If the circuit is already complete, write a *C* beside it.

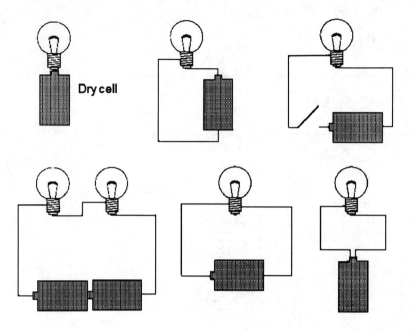

Dry cell

HRW material copyrighted under notice appearing earlier in this work.

214

Answers to Unit 5 End-of-Unit Assessment

Word Usage

1. Answer:
   Sample answer: The *resistance* of the wire in the toaster causes the heating element to heat up.
   Sample answer: The meter reading showed the amount of electrical energy consumed in *kilowatt-hours*.

Correction/Completion

2. Answer:
   a. Sample answer: A battery consists of a *number of cells connected together*.
   b. Sample answer: Energy flow *changes direction* in an alternating current generator.
   c. Sample answer: A *parallel* circuit has at least two branches, with the current divided between each.

Short Response

3. Answer:
   safety pin
   scissors
   nail

4. Answer:
   c__ a unit of one cycle per second
   d__ a unit of measure of electrical charge
   a__ 1 C passing a point in 1 s
   e__ the energy of the charge flowing in a circuit
   b__ the rate at which electrical energy is used
   f__ flow of charge in a conductor

Illustrative

5. Answer:
   a. There needs to be a moving magnet inside the coil.
   b. An energy source, such as a cell or a collection of cells, needs to be added. Also, the switch must be closed in order for an electrical current to flow through the circuit.

6. Answer:
   a. Circuit 1 is a series circuit. Circuit 2 is a parallel circuit.
   b. Circuit 2
   c. The bulbs glow equally bright.
   d. The bulbs glow equally bright.

Numerical Problem

7. Answer:
   Yes. 1500 W + 800 W + 450 W = 2750 W;
   2750 J/s + 110 J/C = 25 C/s, or 25 A, required

   Difficulty: 3

Graphic

8. Answer:
   a. C (ordinary cell)
   b. About 42 hours
   c. A (alkaline cell)

   Difficulty: 3

Illustrative

9. Answer: Sample illustration:

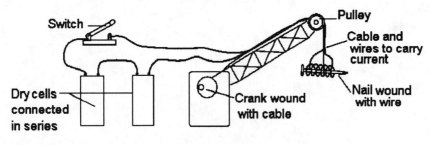

   Difficulty: 4

10. Answer:

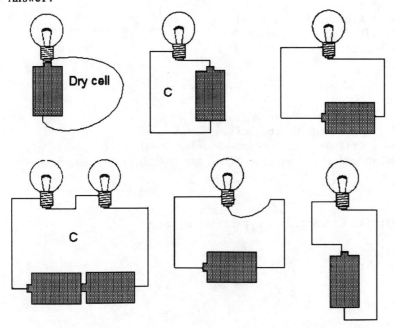

   Difficulty: 4

Activity Assessment

1. Current Action
Teacher's Notes

Overview
Students construct and modify a circuit in order to observe how a change in the direction of electron flow results in a change in the magnetic field.

Materials
(per activity station)
- a strip of aluminum foil (25 cm x 1 cm)
- a D-cell with holder
- 2 lengths of insulated wire (50 cm each)
- a small paper plate
- 2 paper clips
- a magnet

Preparation
Prior to the assessment, equip student activity stations with the materials needed. Remind students that while electricity can be dangerous, the amounts of electricity used in this activity are quite small and unlikely to cause harm.

Time Required
Each student should have 25 minutes at the activity station and 15 minutes to complete the Data Chart.

Performance
At the end of the assessment, students should turn in the following:
- a completed Data Chart

Evaluation
The following is a recommended breakdown for evaluation of this Activity Assessment:
- 40% use of materials to set up a circuit
- 60% ability to make observations and correctly interpret results

2. Current Action

In this activity you'll investigate magnetic fields in two different types of circuits.

Before You Begin . . .

As you work through the tasks, keep in mind that your teacher will be observing the following:
* how you use the materials to set up a circuit
* how well you make observations and correctly interpret results

Go with the flow!

Task 1: Begin by touching the magnet to the foil. What happens? Record your observation and an explanation in the Data Chart on the next page.

Task 2: Place the magnet flat on the plate. Position the foil over the magnet and secure it to the sides of the plate with paper clips, as shown in the diagram. Connect the wires to the D-cell, and then touch the ends of the wire to the paper clips.
Caution: Do not hold the connection for more than a couple of seconds, because the wire will become warm, and the D-cell will drain.
What happens? Record your observations and an explanation in the Data Chart on the next page.

Task 3: Change the direction of electron flow through the foil by reversing the wires to each paper clip. What happens now? Note your observations and an explanation in your Data Chart.

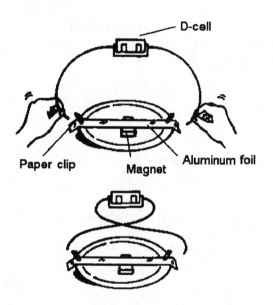

Paper clip   Magnet   Aluminum foil

3. Task 1

Observation	Explanation

Task 2

Observation	Explanation

Task 3

Observation	Explanation

Answers to Unit 5 Activity Assessment

Activity Assessment

1. Answer: Not applicable (teacher's notes)

2. Answer: Not applicable (student's notes)

3. Answer:

Task 1

Observation	Explanation
Nothing happens.	Aluminum is not magnetic.

Task 2

Observation	Explanation
The strip of foil is pulled toward the magnet.	The current is generating a magnetic field passing through the foil.

Task 3

Observation	Explanation
The strip of foil is repelled by the magnet.	The reversal of current causes the magnet to repel the current passing through the foil.

Unit 5 SourceBook Assessment

Multiple Choice

1. If a neutral object loses electrons, it will

   a. become positively charged.   b. become negatively charged.
   c. remain neutral.

2. Which of the following would result in a static charge?

   a. electrons attracting one another
   b. like charges repelling one another
   c. electrons collecting on the surface of an object
   d. electrons moving through a conductor

True/False

3. Only the distance between two charged objects affects the strength of an electric force.

   a. true   b. false

Multiple Choice

4. According to Coulomb's law, if the distance between two charged objects is decreased, the electric force will

   a. remain the same.   b. decrease.   c. increase.

5. Voltage is a measure of

   a. current.   b. potential difference.
   c. wattage.   d. magnetism.

6. The filament in a light bulb gives off heat in addition to light because of the _____ in the filament.

   a. potential difference   b. pressure
   c. voltage   d. resistance

7. Ohm's law states that the _____ increases with increasing _____ but decreases with increasing resistance.

   a. voltage, power
   b. static electricity, amperage
   c. electric current, voltage
   d. potential difference, electric current

True/False

8. Magnetism is not evenly distributed in a magnet.
   a. true   b. false

9. When you cut a bar magnet in half, one half acts as the south pole of the magnet and one half acts as the north pole of the magnet.

   a. true   b. false

Multiple Choice

10. Magnetic lines of force

    a. often cross each other.
    b. are straight lines at the end of a magnet.
    c. have no effect on a compass needle.
    d. can be used to predict which way a compass needle will point.

11. In an atom, magnetic force is caused by

    a. the interaction of protons and electrons.
    b. the motion of unpaired electrons.
    c. the number of paired electrons in the atoms.
    d. electrons orbiting the nucleus.

12. Magnetic domains are _____ in an unmagnetized iron bar.

    a. absent
    b. randomly arranged
    c. aligned
    d. None of the above

13. Which material is naturally magnetic?

    a. alnico   b. magnequench   c. magnetite   d. iron

14. Electromagnets are _____ magnets.

    a. temporary   b. permanent

Short Response

15. When using a compass, you are working with two magnets. What two magnets are they?

    _____

Multiple Choice

16. Which is not involved in the production of alternating current?

    a. brush
    b. commutator
    c. armature
    d. magnetic field

17. A transformer with two or more turns in the secondary coil than in the primary coil

   a. is a step-up transformer.
   b. will decrease voltage.
   c. cannot produce current.
   d. Both a and c

True/False

18. A motor can be thought of as the reverse of a generator.

   a. true   b. false

Matching

19. Match each property on the left with the correct type of force on the right. You may need to use an answer more than once.

   _____ works at a distance   a. electric

   _____ becomes stronger as distance between
           involved objects increases   b. magnetic

   _____ related to electrons   c. both a and b

   _____ determined by amount of charge   d. neither a nor b

Short Response

20. How could hitting a temporary magnet cause it to lose its magnetic properties?

   _____

   _____

   _____

21. Explain how a spark can be released when you walk on a rug and then touch a doorknob.

   _____

   _____

   _____

   _____

22. Distinguish between a step-down transformer and a step-up transformer in terms of voltage.

   _____

   _____

   _____

23. List two of the four things you must have in order to use electricity.

_____

_____

_____

_____

24. How much current (in amps) is needed to light a bulb that has 0.5 Ω of resistance using a 1.5 V battery? (Show your work.)

_____

25. What is the voltage of the current leaving the transformer in the situation described below? Show your work.

turns in the primary coil (T_P) = 120
turns in the secondary coil (T_S) = 480
voltage in the primary coil (E_P) = 40 V

_____

_____

Answers to Unit 5 SourceBook Assessment

Multiple Choice

1. Answer: a. become positively charged.

2. Answer: c. electrons collecting on the surface of an object

True/False

3. Answer: b. false

Multiple Choice

4. Answer: c. increase.

5. Answer: b. potential difference.

6. Answer: d. resistance

7. Answer: c. electric current, voltage

True/False

8. Answer: a. true

9. Answer: b. false

Multiple Choice

10. Answer: d. can be used to predict which way a compass needle will point.

11. Answer: b. the motion of unpaired electrons.

12. Answer: b. randomly arranged

13. Answer: c. magnetite

14. Answer: a. temporary

Short Response

15. Answer: The compass needle and the Earth

Multiple Choice

16. Answer: b. commutator

17. Answer: a. is a step-up transformer.

True/False

18. Answer: a. true

Matching

19. Answer:
    c   works at a distance
    d   becomes stronger as distance between involved objects increases
    c   related to electrons
    a   determined by amount of charge

Short Response

20. Answer:
    It could cause the magnetic domains to become randomly arranged, which would cause them to cancel each other's magnetic properties.

21. Answer:
    When you walk on a rug, your feet may pick up electrons from the atoms of the carpet. This gives your body a negative electric charge. When you get near an uncharged object, like the doorknob, extra electrons flow from you to the object and cause a small spark.

22. Answer:
    A step-down transformer reduces the voltage across a circuit, and a step-up transformer increases the voltage across a circuit.

23. Answer:
    Answers should include two of the following: a source of electricity, a way to manipulate voltage, a device for converting electricity into useful work, and a pathway for the current.

24. Answer: $I = E/R$; $I = 1.5$ V$/0.5$ $\Omega = 3$ A

25. Answer:
    $E_P/E_S = T_P/T_S$; 40 V$/E_S$ = 120 turns/480 turns; E_S = (40 V x 480 turns)/120 turns; E_S = 160 V

Unit 5 Extra Assessment Items

Word Usage

1. The following units of measurement are used frequently in connection with electromagnetic systems: *coulombs, amperes, joules, volts, watts, kilowatts,* and *kilowatt-hours.* Show your understanding of these terms by using each of them in a description of the electrical energy that is used to light a lamp in your home.

   _____

   _____

   _____

   _____

   _____

2. Compare a lead storage battery with a simple dry cell. Use the following words and phrases in your comparison: *electrolyte, number of electrodes, size of the current,* and *ability to be recharged.*

   _____

   _____

   _____

   _____

   _____

   _____

   _____

   _____

   _____

   _____

Correction/Completion

3. Correct the statements below.

   a. Once used up, dry cells can be recharged.

   _____

   _____

   b. A voltmeter measures electrical current. A galvanometer measures electrical potential.

   _____

   _____

   _____

4. Hector has a habit of racing from one idea to the next when he speaks. As a result he often does not finish his sentences. Help Hector by completing his sentences for him.

   "When you think about it, the flow of water in a hose is a lot like

   the flow of electricity in a wire," he began. "Water pressure is a lot

   like _____, and the amount of water flowing through the hose

   is a lot like _____. The size of the hose makes a

   difference, too. A hose's size or shape can restrict the flow of water

   through it. This is a lot like _____."

5. Place the following words in the appropriate blanks in the paragraph: *electrical, chemical, kinetic, heat,* and *light.* One word will be used twice.

   Electrical energy can be produced from _____ energy in a dry

   cell, from _____ energy in a solar cell, from _____

   energy in a thermocouple, and from _____ energy by moving a

   magnet near a coil of wire. An energy change also takes place in a

   wire that resists the flow of current. This change is from

   _____ to _____ energy.

Short Essay

6. Describe how you could use a bar magnet to generate an electric current in a conducting wire. (You can support your description with a sketch, if you like).

_____

_____

_____

_____

_____

_____

_____

_____

_____

7. Sara explained parallel and series circuits using a car to represent the electrons, streets to represent the conducting pathways, the width of the streets to represent resistance, and a roadblock to represent a broken circuit. What might she have said to compare the two types of circuits?

_____

_____

_____

_____

_____

_____

_____

_____

_____

_____

_____

8. Using the theory of charged particles, Peter explained to Bill how a dry cell and copper wire can be used to light a bulb. What might Peter have said?

_____

_____

_____

_____

_____

_____

_____

_____

_____

_____

Short Response

9. Because the wires that conduct a current to an electric lamp carry the same current as the lamp, why don't the wires get as hot as the filament? Circle the best answer from the list below.

   a. They are insulated.
   b. They are longer.
   c. They are thinner.
   d. They have less resistance to the current.

10. Correctly label the diagrams using the following terms: *wet cell, negative electrode, positive electrode, electrolyte,* and *dry cell.* You may need to use some of these terms more than once.

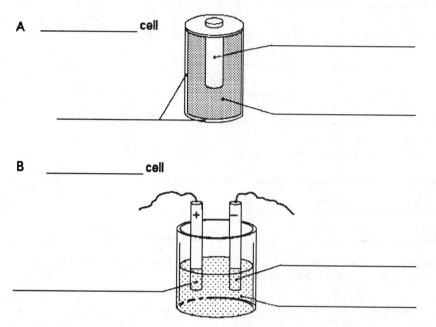

A _____ cell

B _____ cell

11. If the ends of a circuit tester are placed at *A* and *B*, the bulb lights up. The bulb also lights up if the ends are placed at *C* and *D*. But the bulb does not light up if the ends of a circuit tester are placed at *B* and *E*. Find one solution to this puzzle by drawing in connecting wires between the points that would give this result.

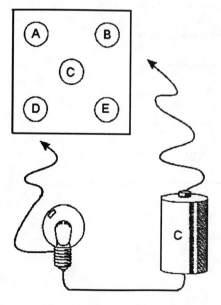

Illustrative

12. Below are a number of diagrams showing how electrical energy is produced. In each case, explain how the current in the wire could be increased.

a.

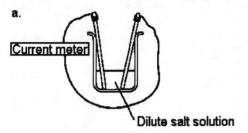

Dilute salt solution

b.

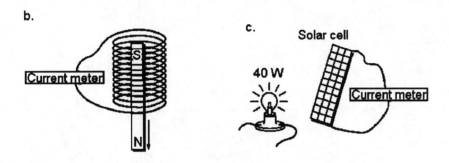

HRW material copyrighted under notice appearing earlier in this work.

233

13. In the illustration below, find as many things as you can that make use of (a) electricity and (b) magnetism. Record your explanations in the table that follows.

Appliance	Uses electricity	Uses magnetism	Explanation

14. In each of the following cases, correct the illustration.

a.

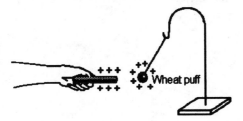

b.

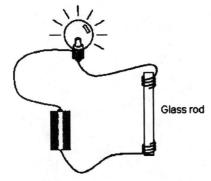

Glass rod

15. Suppose that you have a horseshoe or bar magnet, a galvanometer, a wire coil, and a windmill. Given those materials, design an electric generator. Show your design by means of a diagram.

16. Draw a circuit with two dry cells, a switch, and three light bulbs. If any of the bulbs burn out, the circuit must still work.

17. Wheat puffs are dangling at the end of two threads as shown. Redraw this sketch to show wheat puffs *A* and *B* in each of the situations listed below.

   a. Puff *A* is positively charged.
      Puff *B* is uncharged.
   b. Puff *A* remains positively charged.
      Some negative charge is given to puff *B*.
   c. Puff *A* remains positively charged.
      Puff *B* is also positively charged.
   d. All charge is removed from puff *A*.
      Puff *B* remains positively charged.

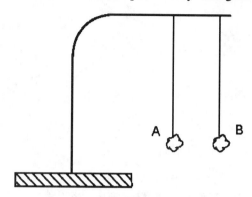

18. Draw a circuit diagram to represent a battery in series with two light bulbs.

19. Suppose that you have two D-cells and two light bulbs.

   a. Draw a circuit diagram with these components to show how you would achieve maximum brightness.

   b. Draw a circuit diagram with these components to show how you would use them so that the D-cells would last as long as possible.

Numerical Problem

20. An electric heater is rated at 480 W.

    a. How many amperes flow through the heater on a 120 V line?

    _____

    _____

    b. How much does it cost to operate the heater for 5 hours if electricity costs 6¢ per kilowatt-hour?

    _____

    _____

21. How many amperes of current will a 720 W electric iron draw from a 120 V line? How many coulombs of charge flow into the iron in a minute?

    _____

    _____

    _____

Performance Task

Assemble the items listed below and conduct the demonstrations as if you were teaching younger students the things that you have learned in this unit.

22. **Materials:** plastic ruler, plastic wrap, vinyl strip, flannel cloth, wheat puff, thread, a stand for suspending the wheat puff (This can be constructed from a standard coat hanger.)

    Tasks:
    a. Demonstrate the charging of a material by friction.
    b. Demonstrate that there is more than one kind of electrical charge.

23. **Materials:** magnetic compass, insulated electrical wire, D-cell, switch

    Task: Demonstrate that a current flowing in a wire has magnetic properties.

24. **Materials:** galvanometer, lemon, copper strip, zinc strip, insulated electrical wire

    Task: Construct a chemical cell and demonstrate that current flows in a closed circuit.

25. **Materials:** insulated electrical wire, cylindrical or bar magnet, bathroom-tissue tube, homemade galvanometer

    Tasks:
    a. Demonstrate the production of alternating electrical current.
    b. Show one way to increase the amount of current.

26. **Materials:** nail, insulated electrical wire, D-cell, masking tape, paper clips

    Tasks:
    a. Construct and demonstrate the operation of an electromagnet.
    b. Using the materials you have at hand, demonstrate one way to increase the electromagnet's strength.

27. Suppose that your job is to design electrical circuits to perform certain tasks. You can develop your skills as a circuit designer with the following materials: 2 D-cells, 4 flashlight bulbs, electrical wire, 2 switches.

   Tasks:
   a. Draw a simple circuit containing a single D-cell and a single flashlight bulb.
   b. Draw a circuit containing 2 D-cells and 4 flashlight bulbs to accomplish the following:
      (i) The useful life of the D-cells is conserved.
      (ii) If any one of the flashlight bulbs burn out, the other bulbs will continue to function.
   c. Now construct this circuit to test it out.

28. Demonstrate and reinforce your understanding of electrical circuits by acting out a performance of their components under different arrangements. You should have at least four actors, two to represent batteries and two to represent light bulbs. Link the components to form circuits by joining hands and feet. Act out the effects of placing the batteries and bulbs in various series and parallel arrangements.

Answers to Unit 5 Extra Assessment Items

Word Usage

1. Answer:
   An acceptable answer uses four of the terms correctly. If all seven terms are used correctly in a way that demonstrates an understanding of each of the terms, give the answer a bonus.

2. Answer:
   Sample answer: The *electrolyte* of a lead storage battery is a liquid (sulfuric acid), while the electrolyte of a dry cell is a paste (ammonium chloride). In a lead battery the *number of electrodes* depends on the number of cells. For example, a lead battery with 6 individual cells will have 14 electrodes—two for each cell and two main electrodes. In a dry cell there are two electrodes. The *size of the current* produced by a lead storage battery can be much greater than that produced by a dry cell. Unlike the dry cell, the lead storage battery has the *ability to be recharged*.

Correction/Completion

3. Answer:
   a. Once used up, dry cells *cannot* be recharged.
   b. A voltmeter measures electrical *potential*. A galvanometer measures electrical *current*.

4. Answer:
   "When you think about it, the flow of water in a hose is a lot like

   the flow of electricity in a wire," he began. "Water pressure is a lot

   like <u>voltage</u>, and the amount of water flowing through the hose is a

   lot like <u>current or amperage</u>. The size of the hose makes a difference,

   too. A hose's size or shape can restrict the flow of water through it.

   This is a lot like <u>resistance</u>."

5. Answer:
   Electrical energy can be produced from <u>chemical</u> energy in a dry

   cell, from <u>light</u> energy in a solar cell, from <u>heat</u>

   energy in a thermocouple, and from <u>kinetic</u> energy by moving a

   magnet near a coil of wire. An energy change also takes place in a

   wire that resists the flow of current. This change is from

   <u>electrical</u> to <u>heat</u> energy.

Short Essay

6. Answer:
An acceptable answer should contain the following two key ideas: (1) A magnet must be moving in relation to a conducting wire. (2) The conducting wire must be part of a closed circuit through which electricity can flow. These ideas can be represented in words, in a diagram, or both. An effective way to generate a current using a wire and a magnet is to form the wire into a coil and move the magnet back and forth inside the coil. Give the answer a bonus if it clearly describes such an arrangement.

7. Answer:
Sample answer: Electrons flowing in the pathways of an electric circuit are like cars moving along a street. When the street is wide, many cars can move through at one time. This is like a circuit with low resistance. When the street is narrow, fewer cars can pass through. This is like a circuit with high resistance. When a street is blocked, the cars cannot move through, so there will be a traffic jam. This is like a circuit that is broken by opening a switch so that the electrons cannot go through. When all of the cars follow the same route through the streets, this is like a series circuit. On the other hand, when some of the cars turn off one street onto other streets, this is like a parallel circuit.

8. Answer:
Sample answer: In a dry cell, electrons build up on the negative electrode. If you connect the negative electrode to the positive electrode with a connecting path such as a copper wire, electrons will flow continuously. If a bulb is placed in the connecting path, the electrons will pass through the filament of the bulb. The filament has a high resistance to the flow of electrons, so it heats up and gives off light.

Short Response

9. Answer: d. They have less resistance to the current.

10. Answer:

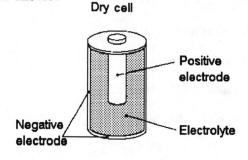

Dry cell

Positive electrode

Negative electrode

Electrolyte

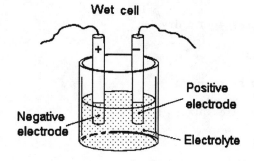

Wet cell

Positive electrode

Negative electrode

Electrolyte

11. Answer: Answers may vary, but the simplest solution is as follows:

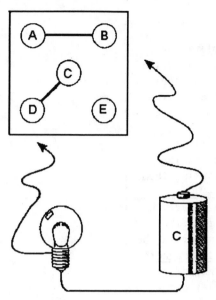

Illustrative

12. Answer:
To increase the current you could do the following:
a. Use a stronger (more concentrated) salt solution; use a different, stronger electrolyte; use larger electrodes.
b. Add more turns to the coil; use a larger, more powerful magnet; move the magnet faster.
c. Use a brighter light; add more solar cells to the circuit.

13. Answer:
    Sample answer:

Appliance	Uses electricity	Uses magnetism	Explanation
refrigerator	X	X	Electricity powers the compressor and light. The motor that drives the compressor contains magnets.
stove	X		Electricity flows through the heating elements, causing them to heat up.
clock	X	X	Electricity powers the motor that runs the clock. The motor in the clock contains magnets.
electric knife	X	X	The motor that drives the knife uses electricity and contains magnets.
electric light	X		Electricity causes the filament in the light bulb to heat up and glow brightly.
toaster	X		Electricity flows through the heating elements of the toaster, causing them to heat up.
can opener	X	X	Electricity drives the motor that turns the can opener. The motor contains magnets.
flashlight	X		Electricity supplied by dry cells causes the filament in the flashlight bulb to glow.

14. Answer:
a.

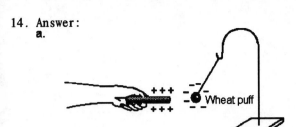

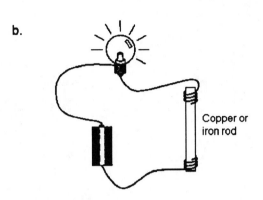

b.

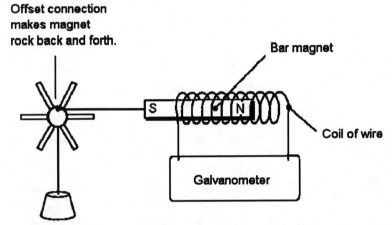

Copper or
iron rod

15. Answer: Sample answer:

**Offset connection
makes magnet
rock back and forth.**

Bar magnet

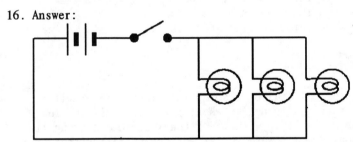

Coil of wire

Galvanometer

16. Answer:

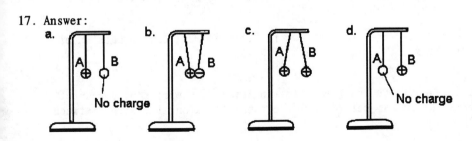

17. Answer:
a.   b.   c.   d.

A | B   A | B   A | B   A | B
No charge   No charge

18. Answer:

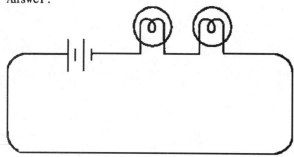

19. Answer:

a

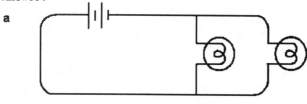

b

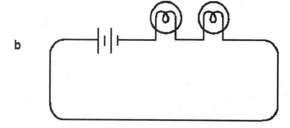

Numerical Problem

20. Answer:
    a. If $W = V \times A$, then 480 W = 120 V x A. Solve for A to get the number of amps. 480 W = 120 V x A; 4 = A. *4 amperes flow through the heater on a 120 V line.*
    b. 0.48 kW x 5 h = 2.4 kWh
       2.4 kWh x 6¢/kWh = 14.4¢, so it would cost about 14¢.

21. Answer:
    If $W = V \times A$, then 720 W = 120 V x A. Solve for A to get the number of amps. 720 W = 120 V x A; A = 6; so 6 amperes flow through the iron on a 120 V line.
    6 A x 60 sec. = 360 C. So 360 coulombs flow into the iron in a minute.

Performance Task

22. Answer:
    Students may demonstrate the charging of a material by rubbing the plastic ruler with plastic wrap and then touching the ruler to the suspended wheat puff. The puff should move away. Students should then rub the vinyl strip with the flannel cloth and bring it near the wheat puff. The wheat puff should move toward the strip, demonstrating that the ruler and the wheat puff are charged objects and that they have opposite charges.

23. Answer:
    If the wire is connected to the D-cell so that a circuit is formed, the compass should move when placed beneath the wire.

24. Answer:
    The copper and zinc strips should be inserted into the lemon. The wire should connect the strips to each other. The galvanometer should be connected along the wire, between the two stripes.

25. Answer:
   Students should wind about 50 turns of wire around the cardboard tube. The ends of the wire should be connected to the galvanometer. When the magnet is moved into and out of the tube, an alternating current is produced in the wire. The amount of current can be increased by either increasing the number of turns of wire or by increasing the speed that the magnet moves into and out of the tube.

26. Answer:
   An electromagnet can be constructed by winding about 20 turns of wire around the nail and connecting the ends of the wire to the D-cell. The nail can then be used to pick up the paper clips. The electromagnet could be strengthened by increasing the number of turns of wire on the nail.

27. Answer:
   a. Circuit diagrams should be clear and logical examples of a simple circuit containing a single D-cell and a single flashlight bulb.
   b. For (i), the bulbs should be connected in a series.
      For (ii), the bulbs should be connected in parallel.

28. Answer:
   Presentations should be clear and should show logical planning as well as understanding of electrical circuits.

Chapter 16 Assessment

Word Usage

1. Using the words *sound*, *amplitude*, *energy*, and *ears*, explain why you think many construction workers wear earplugs when operating jackhammers.

   _____

   _____

   _____

   _____

   _____

Correction/Completion

2. The following sentences are incorrect or incomplete. Your challenge is to make the sentences correct and complete.
   a. The more often an object vibrates within a given period of time, the louder the sound will be.

   _____

   _____

   _____

   b. Words such as *shrill*, *harsh*, and *metallic* refer to the pitch of a sound.

   _____

   _____

   _____

Illustrative

3. Draw a tuning fork that will produce a higher-pitched sound than the one shown below.

Short Essay

4. Do you think people with normal hearing ability ever experience total silence? Explain your answer.

_____

_____

_____

_____

_____

Graphic

5. Use the graph below to answer the question that follows.

On the graph, the bar for dogs appears to be longer than the bar for dolphins or for moths, yet dolphins and moths can hear a much wider range of vibrations than dogs can. How can you explain this?

_____

_____

_____

_____

_____

Sounds Heard by Animals

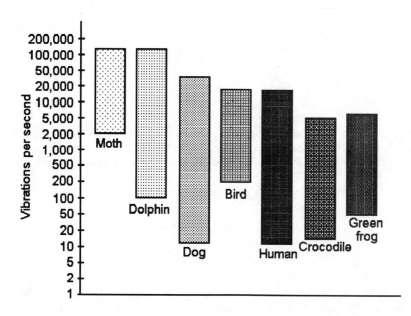

Answers to Chapter 16 Assessment

Word Usage

1. Answer:
   Sample answer: The *sound* produced by jackhammers is very loud because the vibrations they produce have a very large *amplitude*. Construction workers often wear earplugs because the amount of *energy* transferred by these vibrations could damage their *ears*.

Correction/Completion

2. Answer:
   a. The more often an object vibrates within a given period of time, the *higher the pitch of* the sound will be.
   b. Words such as *shrill*, *harsh*, and *metallic* refer to the *quality* of a sound.

Illustrative

3. Answer: Sample drawing:

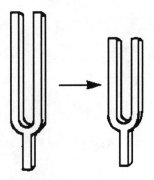

Short Essay

4. Answer:
   Sample answer: No; people who can hear are constantly receiving vibrations, even those that are often "tuned out" and not noticed. Even in a "soundproof" room people hear high-pitched sounds from their nervous system and low-pitched sounds from their circulatory system.

   Difficulty: 3

Graphic

5. Answer:
   Sample answer: The scale on the graph is not set in equal intervals. The bar for dogs is the longest, but it spans only about 50,000 vibrations per second. The bar for dolphins spans about 150,000 vibrations per second, and the bar for moths spans about 147,000 vibrations per second.

   Difficulty: 4

Chapter 17 Assessment

Correction/Completion

1. The following sentences are incorrect or incomplete. Your challenge is to make the sentences correct and complete.

   a. Sound energy can be transmitted through gases, liquids, and outer space.

   _____

   _____

   _____

   b. An object traveling at the speed of sound compresses the air in

   front of it and creates a _____.

   c. Blind people detect the presence of obstacles in their path by sensing pressure on their skin caused by the obstacles.

   _____

   _____

   _____

Short Response

2. a. Number the following materials from 1-4 to show how well they transmit sound. Use 1 for the best transmitter and 4 for the worst.

   cardboard   _____

   cotton   _____

   steel   _____

   wood   _____

   b. What property of a material determines how well it will transmit sound?

   _____

   _____

   _____

3. Nicole wondered how far it was from where she stood in the valley to the mountain wall. She shouted "Hello" and counted 3 seconds before she heard the echo. If sound travels 345 m/s, how far was Nicole from the mountain?

_____

_____

Illustrative

4. The following diagram illustrates a portion of a sound wave in air after the string on a guitar has been plucked, but the artist did not include the labels! Complete the illustration by labeling the following parts: compression regions, expansion regions, and one wavelength.

Short Essay

5. How is the outside of a dog's ear better designed to capture sound than a human ear is?

_____

_____

_____

_____

_____

_____

6. Acoustics is the branch of science that deals with the transmission of sound. It is often a very important consideration for designers of concert halls and auditoriums. Using what you've learned about how sound moves and how sound waves are reflected, sketch a design for an auditorium that would allow the sound from the voice of a single speaker to reach as many areas of the room as possible. You may want to consider building materials as well as the shape of the room.

_____

_____

_____

_____

_____

_____

Answers to Chapter 17 Assessment

Correction/Completion

1. Answer:
   a. Sound energy can be transmitted through gases, liquids, and *solids*. *Sound cannot be transmitted through outer space because sound must have a medium to travel through.*
   b. sonic boom
   c. Blind people detect the presence of obstacles in their path by *hearing the sound waves that bounce off* the obstacles.

Short Response

2. Answer:
   a. cardboard   3
       cotton   4
       steel   1
       wood   2
   b. In general, a material's density, or how close together its particles are, determines how well it will transmit sound.

Numerical Problem

3. Answer: 517.5 m (345 m/s x 3 s = 1035 m; 1035 m ÷ 2 = 517.5 m)

Illustrative

4. Answer:
   Sample answer: (Note: One wavelength is the distance between any two corresponding points on successive waves.)

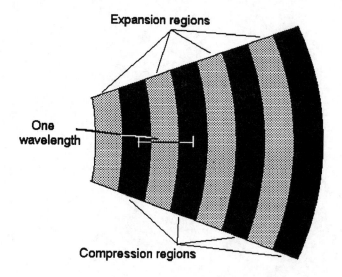

Short Essay

5. Answer:
Sample answer: A dog's ears are better receptors because they are larger than are a human's ears and therefore have more surface area to catch and redirect sound waves. A dog can also move its ears in different directions in order to capture sounds coming from all areas.

Difficulty: 3

Illustrative

6. Answer:
Sketches will vary from simple to elaborate but should show an awareness of how sound waves are reflected by some objects better than by others. Students may sketch an auditorium with seats on an incline or in a circle. They may suggest using panels on the ceiling to help reflect sound waves to specific locations, and they may describe how walls made from wood would help to absorb extra, unwanted reverberation that other materials, such as metal, would transmit.

Difficulty: 4

Chapter 18 Assessment

Word Usage

1. a. Describe how we can "see" sounds by writing a sentence that includes the following words: *fluctuations, sound waves, electric current, oscilloscope,* and *visual image.*

   _____

   _____

   _____

   b. Use the words *pitch, tension,* and *frequency* to explain the following statement: You move pegs to tune the guitar strings.

   _____

   _____

   _____

Numerical Problem

2. John plays the note A on the flute. The air column vibrates 880 times per second.

   a. What would be the frequency if John played the A one octave below this one?

   _____

   _____

   _____

   b. Pam, who is listening to John play the flute, moves twice as far away and listens as John plays the note a second time. Which of the following does she hear?

   • a note vibrating at 1760 Hz
   • a note vibrating at 440 Hz
   • a note twice as loud
   • a note one-half as loud
   • a note one-quarter as loud

   Explain your reasoning.

   _____

   _____

   _____

3. Here are representations of the oscilloscope screens for certain sounds. Five of the sounds were made by tuning forks. The other two were produced by other sound sources. The oscilloscope on which these sound representations appear has a time sweep of 1/100 sec.

   a. Which pictures represent the loudest sounds? the softest sounds?

_____

_____

_____

   b. Which two pictures show two sounds of the same loudness but an octave apart?

_____

_____

_____

   c. What is the frequency of the sound pictured in *B?*

_____

_____

_____

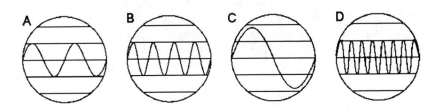

4. A violin string, when bowed, vibrates in more than one way at the same time. Illustrate this with a drawing.

Data for Interpretation

5. Use the table below to answer the questions that follow.

Change in sound energy	Increase in loudness
2X	+3 dB
4X	+6 dB
8X	+9 dB
10X	+10 dB
100X	+20 dB
1000X	+30 dB

a. By how much must sound energy be increased to produce a 12 dB increase in loudness?

_____

_____

_____

b. A certain sound has a loudness of 6 dB. Another sound has 800 times as much sound energy as the first. How loud is the second sound? Show your work.

_____

_____

_____

c. Your friend Kiki tells you about the last music concert she attended and says, "It couldn't have been that dangerously loud. After all, ordinary conversation has a loudness of 60 dB, and a rock concert is only twice as many decibels, so a concert is only as damaging as hearing two conversations at once." What would you say to her?

_____

_____

_____

_____

_____

_____

_____

_____

_____

Answers to Chapter 18 Assessment

Word Usage

1. Answer:
   a. Sample answer: An *oscilloscope* changes *soundwaves* into *fluctuations* in an *electric current* to produce a *visual image* of sounds.
   b. Sample answer: The pegs in a guitar are wound to change the *tension* of the guitar strings and hence the *frequency* at which they vibrate. The result is a change in *pitch* of the tone produced.

Numerical Problem

2. Answer:
   a. One octave below a given note corresponds to one-half of the given frequency, so the new frequency would be 440 Hz.
   b. • a note one-quarter as loud
      At twice the distance, the sound wave would cover four times the area and would be one-quarter as loud as it sounded at Pam's original distance from the flute.

Illustrative

3. Answer:
   a. *E* and *F* have the same amplitude and are the loudest sounds. *A*, *B*, and *D* have the same amplitude and are the softest sounds.
   b. *A* and *B*, with *B* being the higher pitch
   c. Four wavelengths in 1/100 second corresponds to a frequency of 400 wavelengths in one second, or 400 Hz.

4. Answer: Sample answer:

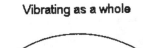

Vibrating as a whole

Vibrating as a whole and in two parts at the same time

Difficulty: 3

5. Answer:
   a. Because 12 = 3 + 9, sound energy must be increased by 8 x 2, or 16.
   b. 800, or 8 x 100, times as much sound energy corresponds to a 29 dB, or 20 dB + 9 dB, increase in loudness, for a total loudness of 6 dB + 29 dB, or 35 dB.
   c. Answers will vary. Sample answer: Kiki is correct in estimating the loudness of a rock concert at 120 dB. The fact that this is twice as many decibels as a normal conversation, however, does not mean that it is like hearing two conversations at once. An additional 60 dB of loudness corresponds to an increase of 1,000,000 in sound energy. This greater amount of sound energy could be very dangerous to the middle ear.

Difficulty: 4

Unit 6 End-of-Unit Assessment

Word Usage

1. In one or more sentences, explain how the "boom" of a bass drum reaches your ears. In your explanation, use the words *vibration*, *compression*, and *expansion*.

_____

_____

_____

_____

Correction/Completion

2. The following statements are incorrect or incomplete. Your challenge is to make them correct and complete.

   a. When more energy is added to an object, it vibrates with less amplitude, and the sound generated is softer.

   _____

   _____

   _____

   b. The speed of sound can be found by multiplying the distance that a sound travels by the time it takes to travel that distance.

   _____

   _____

   _____

   c. If two objects are vibrating at the same frequency (their natural frequency), the sound will be softer.

   _____

   _____

   d. A _____ guitar string vibrates at a pitch higher than

   that of a _____ guitar string.

Short Response

3. Describe how you could make each of the following objects vibrate faster:

   a. a ruler

   _____

   _____

   b. a rubber band

   _____

   _____

4. Identify the following for each sound listed below: high, low, or mixed frequencies, large or small amplitudes, and whether the sound is a complex or a simple vibration.

   a. A small tuning fork is struck hard.

   _____

   _____

   b. A violinist slides her finger down the string and plays a note softly.

   _____

   _____

5. Match the descriptions on the right with the corresponding effects on the left.

   a. echolocation

   b. forced vibration
   increases loudness

   c. concentration of
   sound energy in a
   specific direction

   d. resonance

   _____ An opera singer sings a high note and a crystal glass shatters.

   _____ Mossimo cups his hands around his mouth to call to a friend across the street.

   _____ Jeannie, a blind woman, stops in front of a table placed in her path.

   _____ A turning fork placed on an overturned metal bucket makes a loud noise.

Illustrative

6. The diagrams below show four different situations involving pendulums.

   Compare the frequency of the vibration of the following:

   a. pendulum *A* and pendulum *B*

   _____

   _____

   b. pendulum *C* and pendulum *D*

   _____

   _____

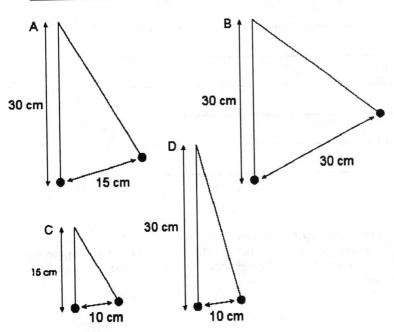

Numerical Problem

7. Barrett is standing by a lake. Across the lake is a cliff. He calls out his name, and 2.6 seconds later he hears an echo. If the sound travels at the rate of 340 m/s, how far is the cliff from Barrett? Show your work.

   _____

   _____

   _____

   _____

   _____

8. How much more sound energy does it take to change a whispered conversation (30 dB) into an ordinary conversation (60 dB)? (Remember that with each increase of 10 dB, the sound energy increases by a factor of 10.) Show your work.

_____

_____

_____

_____

Short Essay

9. Your friend has heard about something called a sonogram, which physicians use to examine a fetus inside the womb. Your friend says, "If it has to do with sound, it must hear and measure the heartbeat." How would you answer your friend?

_____

_____

_____

_____

_____

Illustrative

10. The circles in the diagram below represent sound coming from the engines of a stationary airplane, which is represented by x. How would you redraw the diagram to indicate the sound waves as the plane flies through the air at a speed
   a. less than the speed of sound?
   b. equal to the speed of sound?

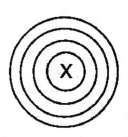

Word Usage

11. The following story explains what happens when you hear a door slam, but the chapters are out of order. Use the following words to complete the story: *cochlea, eardrum, brain, pinna, stirrup, hair cells, ear canal, anvil, oval window, auditory nerve, stirrup,* and *nerve cells.* Then, in the space above each paragraph, write chapter numbers to arrange the sections in order.

_____ Vibrations in the fluid cause _____ to vibrate. This stim- ulates _____ to send out impulses.	_____ The stirrup vibrates against the _____ of the inner ear. The vibra- tions are transferred to the fluid-filled _____.	_____ A door slams. This causes air particles to vibrate. The sound moves through the air in waves of compression and expansion.
_____ Next the _____ vibrates, causing three bones in the middle ear, called the _____, _____, and _____, to vibrate.	_____ Impulses travel along the large _____ _____ until they reach the _____, which interprets the impulses as sound.	_____ The _____ of the outer ear catches the sound waves. Air particles inside the _____ _____ begin to vibrate.

Short Essay

12. Sound travels more quickly through water than through air, but it travels more quickly through warm air than through cold air. Explain why this makes sense.

_____

_____

_____

_____

_____

_____

_____

_____

_____

Answers to Unit 6 End-of-Unit Assessment

Word Usage

1. Answer:
   Sample answer: The *vibration* of the bass drum causes the air around it to vibrate. The particles of air vibrate in a series of *compressions* and *expansions*. When these vibrations reach your ears, you hear the sound.

Correction/Completion

2. Answer:
   a. When more energy is added to an object, it vibrates with *greater* amplitude, and the sound generated is *louder*.
   b. The speed of sound can be found by *dividing* the distance that a sound travels by the time it takes to travel that distance.
   c. If two objects are vibrating at the same frequency (their natural frequency), the sound will be *louder*.
   d. short
      long

Short Response

3. Answer:
   a. Sample answers: Shorten the ruler; make the ruler narrower.
   b. Sample answers: Tighten the rubber band; reduce its width or its length.

4. Answer:
   a. High frequency, large amplitude, simple vibration
   b. Mixed frequency, small amplitude, complex vibration

5. Answer:
   d̲ An opera singer sings a high note and a crystal glass shatters.

   c̲ Mossimo cups his hands around his mouth to call to a friend

   across the street.

   a̲ Jeannie, a blind woman, stops in front of a table placed in her

   path.

   b̲ A turning fork placed on an overturned metal bucket makes a loud

   noise.

Illustrative

6. Answer:
   a. The frequencies are the same.
   b. Pendulum *C* has a higher frequency than pendulum *D*.

Numerical Problem

7. Answer:
The distance to the cliff is one-half of the time multiplied by the speed of sound, so the cliff is 1.3 s x 340 m/s, or 442 m, away.

8. Answer:
The loudness increases by 30 dB, or 3 x 10dB, so the sound energy increases by 3 factors of 10, or 1000.

Short Essay

9. Answer:
Sample answer: A sonogram makes use of sound waves, but it does not hear the fetus's heartbeat. Sound waves are directed into a woman's womb, where they bounce off the fetus's tissues. The waves are then converted into electronic signals, which produce a "picture" of the fetus.

Difficulty: 3

Illustrative

10. Answer:

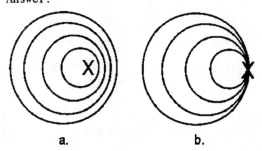

a.   b.

Difficulty: 3

Word Usage

11. Answer:

Chapter 5	Chapter 4	Chapter 1
Vibrations in the fluid cause <u>hair cells</u> to vibrate. This stimulates <u>nerve cells</u> to send out impulses.	The stirrup vibrates against the <u>oval window</u> of the inner ear. The vibrations are transferred to the fluid-filled <u>cochlea</u>.	A door slams. This causes air particles to vibrate. The sound moves through the air in waves of compression and expansion.
Chapter 3	Chapter 6	Chapter 2
Next the <u>eardrum</u> vibrates, causing three bones in the middle ear, called the <u>hammer</u>, <u>anvil</u>, and <u>stirrup</u>, to vibrate.	Impulses travel along the large <u>auditory nerve</u> until they reach the <u>brain</u>, which interprets the impulses as sound.	The <u>pinna</u> of the outer ear catches the sound waves. Air particles inside the <u>ear canal</u> begin to vibrate.

Difficulty: 4

Short Essay

12. Answer:
Students should explain that sound travels better through water than through air because the particles of matter are closer together in liquids than in gases, which makes it easier for sound energy to be transmitted. However, the case is a bit different when comparing cold air and warm air. Although the particles of cold air are closer together than the particles of warm air, sound travels more quickly through warm air because when air is heated, its particles bump into each other more often and are better able to transmit the sound energy.

Difficulty: 4

Unit 6 Activity Assessment

Activity Assessment

1. **Bouncing Off the Walls**
   Teacher's Notes
   Overview
   Students demonstrate how sound bounces off a smooth, hard surface better than off a fabric or foam surface.

   Materials
   (per activity station)
   • 2 stiff cardboard mailing tubes, approximately 50 cm long and 10 cm wide.
   • an alarm clock that ticks
   • a piece of thick, smooth poster board (40 cm x 40 cm)
   • pieces of terry cloth fabric and foam rubber (40 cm x 40 cm or larger)
   • a wall or a stack of books
   • a metric ruler

   Preparation
   Prior to the assessment, equip student activity stations with the materials needed for each experiment.

   Time Required
   Each student should have 15 minutes at the activity station and 25 minutes to complete the Data Chart.

   Performance
   At the end of the assessment, students should turn in the following:
   • a completed Data Chart

   Evaluation
   The following is a recommended breakdown for evaluation of this Activity Assessment:
   • 30% appropriate use of equipment and materials
   • 40% ability to support the hypothesis and to predict appropriate uses of the materials in a concert hall
   • 30% ability to analyze and suggest improvements to the experiment's design

2. Bouncing off the Walls

Some materials strongly absorb sound waves; others strongly reflect them. Acousticians (scientists who study sound) apply these principles to design structures with certain acoustical properties. By combining sound-absorbing surfaces and sound-reflecting surfaces in the right ways, the desired acoustical properties are achieved. In the tasks that follow, you will explore the acoustical properties of certain materials and use your findings to solve a practical problem.

Before You Begin . . .

As you work through the tasks, keep in mind that your teacher will be observing the following:

• how well you use the materials
• your ability to support your hypothesis and to predict appropriate uses of the surfaces for a concert hall
• your ability to demonstrate how different surfaces reflect sounds in different ways

Now you are ready to explore bouncing sounds!

Task 1 Above the Data Chart on the next page, write a hypothesis that states which surfaces will reflect sound the best and the worst.

Task 2 Arrange the materials as shown below and then carry out the experiment. Note: Make sure that your ear is completely within the end of the tube. Cover your other ear with your hand. Record your observations in the Data Chart on the next page.

Task 3 Place a piece of fabric over the posterboard, and repeat Task 2.

Task 4 Rmove the fabric, and cover the posterboard with a piece of foam rubber. Repeat Task 2.

Task 5 Compare the quality of the sound from all three trials. In the Data Chart on the next page, record your conclusions about how sound is reflected by different materials. Do your results support or refute your hypothesis? With which surface did you hear the sound loudest? softest?

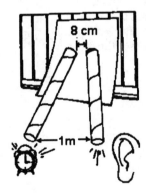

3. Hypothesis

_____

_____

Data Chart

	Observations
Smooth, hard surface	
Fabric-covered surface	
Foam-surface	

Conclusions

_____

_____

_____

_____

Summing Up...

Concert halls present special problems for designers. The designer must strike a
delicate balance in his or her use of reflective and absorbent surfaces.
Usually, the ceilings of concert halls are made to reflect sound, while the
walls, fixtures, and seats are made to absorb sound.

1. Why do you think concert halls are arranged this way?

_____

_____

_____

2. Why would you *not* want the surfaces inside the concert hall to be made either of
   all soft materials or all hard and smooth materials?

_____

_____

_____

Answers to Unit 6 Activity Assessment

Activity Assessment

1. Answer: Not applicable (teacher's notes)

2. Answer: Not applicable (student's notes)

3. Answer:
   Hypothesis
   Answers will vary but should be clear and logical.

Data Chart

	Observations
Smooth, hard surface	Students should observe that the sound of the ticking clock is strongly reflected from the cardboard, as indicated by the sound being clearly audible.
Fabric-covered surface	Students should observe that the cloth-covered surface partially absorbs the sound of the ticking clock, as indicated by the sound being somewhat muffled.
Foam	Students should observe that the foam rubber strongly absorbs the sound of the ticking clock.

Conclusions
Hard, smooth surfaces strongly reflect the sound striking them, whereas soft, uneven surfaces absorb the sound somewhat. Soft, porous materials like foam rubber absorb sound very strongly.

Summing up...
1. The hard ceilings help to transmit the sound to the far parts of the concert hall (such as the balcony) and increase the volume of the sound as the echoes from the ceiling blend with the sound coming directly from the source. The soft surfaces keep the sound from reflecting again and again.
2. If there are too many hard surfaces, the sound will continue to reverberate for a long time, resulting in a confusing muddle of noise. Too many soft surfaces would cause the sound to be absorbed before it could reach the far parts of the hall.

Unit 6 SourceBook Assessment

Multiple Choice

1. Waves transmit

   a. matter.
   b. radiation.
   c. motion.
   d. energy.

2. A wave in which the motion of the particles of the medium is perpendicular to the path of the wave is called

   a. longitudinal.
   b. transverse.
   c. infrasonic.
   d. ultrasonic.

3. As the energy in a wave increases, the amplitude

   a. increases.   b. decreases.   c. stays the same.

True/False

4. You can increase the wavelength of a transverse wave but not a longitudinal wave.

   a. true   b. false

Multiple Choice

5. A wave with a large wavelength will have a relatively _____ frequency.

   a. low   b. high

6. Hertz (Hz) is a measure of

   a. distance between crests.
   b. meters per second.
   c. vibrations per second.
   d. the speed of sound.

7. An echo occurs when a sound wave

   a. is reflected, or bounces off a material.
   b. passes through a vacuum.
   c. turns a corner.
   d. changes direction as its speed changes.

8. When a wave passes from one medium to another, _____ occurs as the speed of the wave changes, and the wave bends.

   a. reflection
   b. refraction
   c. diffraction
   d. interference

True/False

9. Sound travels by longitudinal waves.

   a. true   b. false

10. Only transverse waves can form standing waves.

    a. true   b. false

Multiple Choice

11. The sound waves in front of a moving car seem to be _____ than the sound waves behind the car.

    a. farther apart with a higher frequency
    b. closer together with a higher frequency
    c. farther apart with a lower frequency
    d. closer together with a lower frequency

12. The sound produced by an approaching car has a certain pitch. As the car passes, the pitch changes. Scientists call this phenomenon the _____ effect.

    a. Charles
    b. Coriolis
    c. Doppler
    d. Bernoulli

13. What kind of waves experience the effect described in question 12?

    a. sound waves
    b. light waves
    c. water waves
    d. a, b, and c

14. The P waves of an earthquake are longitudinal waves. Through what medium do these waves move?

    a. the molten center of the Earth
    b. rock and soil
    c. air particles
    d. surface waves

15. The surface waves of an earthquake are caused by

    a. primary waves.
    b. secondary waves.
    c. the interaction of primary and secondary waves with the Earth's surface.

16. Sound waves with frequencies too low for humans to hear are called _____ waves.

    a. infrasonic   b. ultrasonic   c. supersonic   d. neosonic

True/False

17. Hearing damage from loud sounds can almost always be reversed.

    a. true   b. false

Matching

18. Match the descriptions on the left with the wave properties on the right.

    _____ The distance from one crest to the next   a. amplitude

    _____ The rate at which vibrations are produced   b. wave speed

    _____ How fast a wave travels   c. frequency

    _____ The size of a wave   d. wavelength

19. Match the situations on the left with the correct wave characteristic on the right.

    _____ You can hear sound produced around a corner.   a. reflection

    _____ You can't hear the announcer clearly at a football game.   b. diffraction
                                                c. refraction

    _____ A pencil appears bent or broken when it's placed in a glass of water.   d. interference

    _____ You yell and hear an echo.

Short Response

20. Distinguish between constructive interference and destructive interference.

    _____

    _____

    _____

    _____

21. Identify whether the following situations involve infrasonic or ultrasonic waves:

    a. A bat searches for a mosquito to eat. _____

    b. A volcano erupts. _____

    c. Scientists measure the peaks and valleys on the sun's surface.

    _____

    d. A submarine navigates on the ocean floor. _____

22. How does a sonogram capture the image of a fetus inside the mother's womb?

_____

_____

_____

_____

23. If you place a penny in the bottom of a bowl, push the bowl away from you until the penny just disappears from view, and then slowly add water to the bowl, the penny seems to magically reappear. What characteristic of waves might cause this phenomenon? Explain.

_____

_____

_____

_____

24. What is the speed of a wave that has a frequency of 5 Hz and a wavelength of 3 m? Show your work.

_____

_____

25. Draw a diagram of a standing wave that shows one full wavelength with three nodes. Label a node and an antinode.

Answers to Unit 6 SourceBook Assessment

Multiple Choice

1. Answer: d. energy.

2. Answer: b. transverse.

3. Answer: a. increases.

True/False

4. Answer: b. false

Multiple Choice

5. Answer: a. low

6. Answer: c. vibrations per second.

7. Answer: a. is reflected, or bounces off a material.

8. Answer: b. refraction

True/False

9. Answer: a. true

10. Answer: b. false

Multiple Choice

11. Answer: b. closer together with a higher frequency

12. Answer: c. Doppler

13. Answer: d. a, b, and c

14. Answer: b. rock and soil

15. Answer: c. the interaction of primary and secondary waves with the Earth's surface.

16. Answer: a. infrasonic

True/False

17. Answer: b. false

Matching

18. Answer:
    __d__ The distance from one crest to the next
    __c__ The rate at which vibrations are produced
    __b__ How fast a wave travels
    __a__ The size of a wave

19. Answer:
    __b__ You can hear sound produced around a corner.
    __d__ You can't hear the announcer clearly at a football game.
    __c__ A pencil appears bent or broken when it's placed in a glass of water.
    __a__ You yell and hear an echo.

Short Response

20. Answer:
    Constructive interference occurs when the crests and troughs of two waves collide and result in a wave with higher crests and lower troughs. Destructive interference occurs when the crest of one wave collides with the trough of another wave and they cancel each other.

21. Answer:
    a. Ultrasonic
    b. Infrasonic
    c. Infrasonic
    d. Ultrasonic

22. Answer:
    Ultrasonic waves bounce off the fetus inside the mother's body. The waves are converted into electric signals that a computer uses to generate an image.

23. Answer:
    Refraction would cause the light waves to bend at an angle around the edge of the bowl, allowing you to see the penny.

24. Answer: $v = \lambda f$; $v = 3 \times 5$ Hz; $v = 15$ m/s

25. Answer:

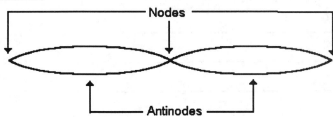

Unit 6 Extra Assessment Items

Word Usage

1. For each statement below, select at least one of the following terms and explain how it relates to the statement: *amplitude, cochlea, decibels, echoes, echolocation, frequency, low-pitched sound, noise, octave, reflection, resonance, semicircular canals, speed of sound, syrinx, tension,* and *timbre.*

a. "Listening to that music makes me dizzy."

_____

_____

_____

_____

_____

b. "That pendulum is swinging farther than this one."

_____

_____

_____

_____

_____

c. "This pendulum is swinging faster than that one."

_____

_____

_____

_____

_____

_____

_____

2. For each statement below, select at least one of the following terms and explain how it relates to the statement: *amplitude, cochlea, decibels, echoes, echolocation, frequency, low-pitched sound, noise, octave, reflection, resonance, semicircular canals, speed of sound, syrinx, tension,* and *timbre.*

a. "The ruler is moving back and forth about 30 times per second."

_____

_____

_____

_____

_____

b. "This tuning fork is vibrating twice as many times per second as the larger one."

_____

_____

_____

_____

c. "When I speak into this metal bucket my voice sounds so full and big."

_____

_____

_____

_____

_____

3. For each statement below, select at least one of the following terms and explain how it relates to the statement: *amplitude, cochlea, decibels, echoes, echolocation, frequency, low-pitched sound, noise, octave, reflection, resonance, semicircular canals, speed of sound, syrinx, tension,* and *timbre.*

a. "My voice sounds the best when I sing in the shower."

_____

_____

_____

_____

b. "That's music I call soft and mellow!"

_____

_____

_____

c. "Has anyone seen my earplugs?"

_____

_____

_____

_____

4. For each statement below, select at least one of the following terms and explain how it relates to the statement: *amplitude, cochlea, decibels, echoes, echolocation, frequency, low-pitched sound, noise, octave, reflection, resonance, semicircular canals, speed of sound, syrinx, tension,* and *timbre.*

a. "How do birds produce those flute-like trills?"

_____

_____

_____

_____

b. "The car sure sounds different after I pass those large buildings and drive past the park."

_____

_____

_____

c. "Bats have no trouble making a big meal of insects in the middle of the night!"

_____

_____

_____

_____

5. Each statement below relates an important idea about sound. For each statement, select one of the following ideas and explain how the idea relates to the statement: *amplitude, cochlea, decibels, echoes, echolocation, frequency, low-pitched sound, noise, octave, reflection, resonance, semicircular canals, speed of sound, syrinx, tension,* and *timbre.*

a. "I wouldn't hear a thing if it weren't for those tiny vibrating hair cells in my ear."

_____

_____

_____

_____

b. "You move these pegs to tune the guitar strings."

_____

_____

_____

_____

c. "It takes only one-half of a second for the sound to travel across the lake. How big do you think the lake is?"

_____

_____

_____

_____

6. Explain the statement below using the following words: *sound, pitch,* and *vibration.*

Cecile plucks the D string of her guitar with one hand while moving the fingers of her other hand along it.

_____

_____

_____

_____

7. Use the words *echolocation* and *frequency* to compare how bats and humans hear.

_____

_____

_____

_____

Correction/Completion

8. Correct the following statement:

   The shorter a pendulum, the greater the time of the vibration and the higher the frequency of the vibration.

   _____

   _____

   _____

   _____

9. Correct the following statement:

   The faster a string vibrates, the louder the generated sound.

   _____

   _____

   _____

   _____

10. Correct the following statement:

    All of these things might produce high-pitched sounds: a very short pipe; a long, tight rubber band; a thin, loose violin string; a large vibrating object; and a small object dropped on the floor.

    _____

    _____

    _____

    _____

11. Correct the following statement:

    The amplitude of a sound from a vibrating object results from the object vibrating in more than one way at a given time.

    _____

    _____

    _____

    _____

Short Essay

12. Sound is produced by vibrating objects. Describe how you would demonstrate or explain this concept to a younger student.

_____

_____

_____

_____

_____

_____

_____

_____

13. Bones conduct sound. Describe how you would demonstrate or explain this concept to a younger student.

_____

_____

_____

_____

_____

14. High sounds are produced by higher frequencies. Describe how you would demonstrate or explain this concept to a younger student.

_____

_____

_____

_____

_____

_____

15. Loud sounds are produced by greater amplitudes. Describe how you would demonstrate or explain this concept to a younger student.

_____

_____

_____

_____

_____

_____

_____

_____

16. Choose any two musical instruments you wish. Describe the differences in their sounds that help you distinguish one from the other. Then describe how you can get higher and louder sounds from each.

_____

_____

_____

_____

_____

_____

_____

_____

_____

_____

_____

17. You want to show someone that you can measure the speed of sound. What would you say or do?

_____

_____

_____

_____

_____

_____

Short Response

18. Sonia is rushing home from school, trying not to get too wet from the rain. In the distance she sees a flash of lightning, and a short time later she hears a roll of thunder. How would you explain to Sonia why she doesn't hear the thunder at the same time that she sees the lightning?

_____

_____

_____

_____

19. For each item below, identify what causes the sound. Tell whether the sound has a high or low frequency and whether the amplitude is large or small.

a. a doorbell

_____

_____

b. a piccolo

_____

_____

c. the roar of Niagara Falls

_____

_____

20. For each item below, identify what causes the sound. Tell whether the sound has a high or low frequency and whether the amplitude is large or small.

a. the bang of a starter pistol

_____

_____

b. a tuba

_____

_____

c. the roar of a race car engine

_____

_____

_____

d. the crash of cymbals

_____

_____

_____

21. In which of the following pairs are the items correctly related? Circle all that apply.

a. long violin string
   low-pitched note
b. high frequency
   low-pitched note
c. thick violin string
   low-pitched note
d. large amplitude
   loud sound
e. short, thin suspended iron pipe
   low-pitched sound
f. long suspended iron pipe
   low-pitched sound

Graphic

22. Jamie wanted to know if there was a relationship between the length of a pendulum and the frequency of its vibration. He designed an experiment and collected the data shown in the table and graph. Use the data to answer the questions that follow.

Length (cm)	Time for 10 vibrations (sec.)	Number of vibrations in 1 sec.
5	4.50	2.20
10	6.50	1.50
15	7.75	1.30
20	8.40	1.10
30	11.00	0.90
35	11.90	0.84
40	12.70	0.79
45	13.50	0.74

a. How should Jamie describe the relationship between the length of a pendulum and its frequency?

_____

_____

_____

b. What length of pendulum would make a good clock?

_____

_____

_____

c. Would you consider Jamie to be a good experimenter? Why or why not?

_____

_____

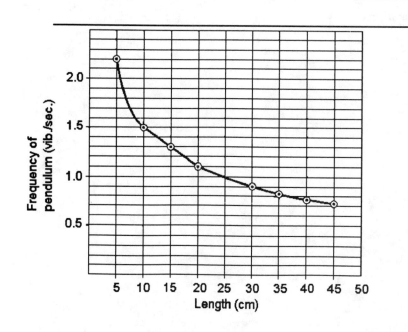

23. Maria is at the end of a long field. She hits a drum loudly two times every second. Five observers spread themselves along the field so that they hear the sounds of the drum either when Maria's hand is up in the air or down striking the drum. They recorded their data in the table below.

Person	Position of Maria's hand	Time for the sound to reach the person (sec.)	Distance from the person to Maria (m)
Phil	up	0.25	85
Dwayne	down	0.50	165
Elsie	up	0.75	260
Vo	down	1.00	335
Leo	down	1.50	500

a. Why did Phil hear the sounds when Maria's hand was up in the air? What is the time between hits?

_____

_____

_____

_____

_____

b. How could the experimenters calculate the time for the sounds to reach them without using stopwatches?

_____

_____

_____

c. Plot the graph of distance and time. Draw a smooth line through the five points.

Does the line appear to pass through the point (0, 0) of the graph? Should it?

_____

d. What was the speed of sound calculated by each observer? What is the average speed of sound according to the observers?

_____

_____

_____

e. Where would you locate yourself between Vo and Leo so that you would hear the drumbeats on Maria's upswing?

_____

_____

_____

| a | b | c | d | e | f |

Maria hits the drum 2 times per second.

Phil hears the sounds only when Maria's hand is up in the air.

Dwayne hears the sounds each time Maria's hand is down, appearing to hit the drum.

Elsie hears the sounds when Maria's hand is up.

Vo hears the sounds when Maria's hand is down.

Leo hears the sounds when Maria's hand is down.

24. The following are sounds as they would be seen on an oscilloscope screen. The first sound is made by a flute playing a note with a frequency of 200 Hz.

a. Which sounds would you classify as musical sounds? as noise?

_____

_____

_____

b. Which is the softest sound? the lowest-pitched sound? the highest-pitched sound?

_____

_____

_____

_____

c. What is the frequency of the sound displayed in *F*?

_____

_____

_____

_____

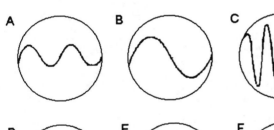

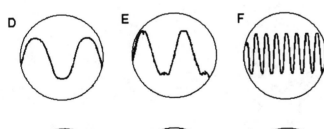

25. In each of the illustrations below, the blade of a saw is clamped to a table. The free end is pulled down as indicated and released.

    a. Which one vibrates the fastest?   _____

    b. Which one vibrates the slowest?   _____

    c. Which one produces the lowest sound?   _____

    d. Which one produces the highest sound?   _____

    e. Which one produces the loudest sound?   _____

a.

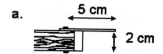

b.

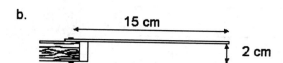

c.

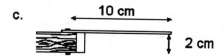

d.

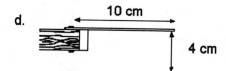

26. Help Liz with her drawing. Find the mistakes she has made. (There are at least eight.)

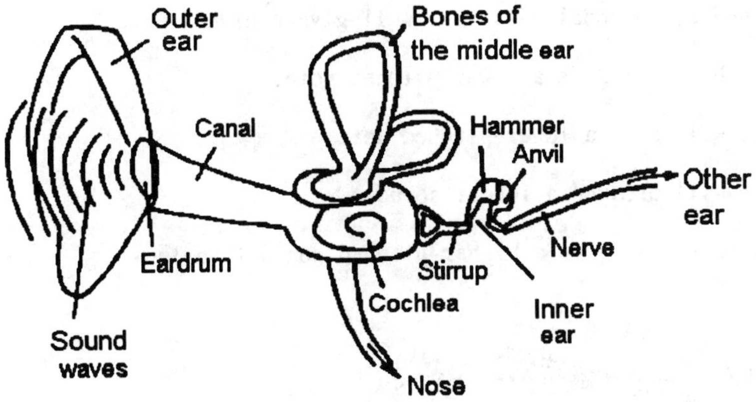

_____

_____

_____

_____

_____

_____

_____

_____

_____

_____

_____

_____

Outer ear

Bones of the middle ear

Canal

Hammer

Anvil

Other ear

Eardrum

Nerve

Stirrup

Cochlea

Inner ear

Sound waves

Nose

27. Matthew drew the sound waves that he thought might be created by two different tuning forks with the same period of vibration (1/10 sec., for example). What is wrong with his drawing?

_____

_____

_____

_____

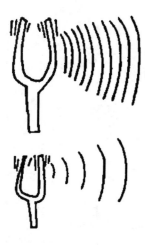

28. For the items below, follow each set of directions.

   a. Draw another string stretched by an equal mass that will give a higher note.

   b. Draw another straw whistle that will give a lower-pitched note.

   c. Draw a measuring stick that will give a lower-pitched note.

   d. Draw a measuring stick that will produce a louder sound.

   e. Draw a set of band cymbals that will give a higher-pitched sound than those pictured.

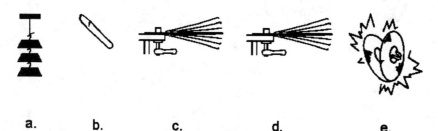

   a.   b.   c.   d.   e.

29. Draw the apparatus that you could use to show the following:

   a. Air moves when sound is transmitted through it.

   b. Sound is transmitted through water.

   c. Sound can be channeled through wood.

   d. A tuning fork sound can be made louder.

Numerical Problem

30. Susan is standing close to the school wall. She claps her hands at the rate of 10 claps in 5 seconds and listens to the echoes. She backs away from the wall until the echoes are covered up by her claps. What distance is Susan from the wall? (Use 340 m/s for the speed of sound.)

_____

_____

_____

_____

Performance Task

31. **Materials:** string, hooked masses
    Task: Determine how the mass of a pendulum bob affects the number of vibrations that the pendulum makes per second.
    a. What is your expected result? That is, what hypothesis are you going to test?

_____

_____

b. What variables should you control (hold constant) in order to do a fair test?

_____

_____

c. Using at least three different hooked masses for the pendulum bob, measure the time for 10 vibrations in order to find the frequency of vibration for each. Record your results here.

_____

_____

d. How do you think experimental errors might affect your results?

_____

_____

e. What conclusions can you make in regard to your initial hypothesis?

_____

_____

_____

32. Design and construct your own musical instrument and explain how it works.

Answers to Unit 6 Extra Assessment Items

Word Usage

1. Answer:
   a. Sample answer: *Amplitude, noise,* and a high *decibel* level are all terms that can be used to describe how music could cause dizziness. This dizziness might be associated with vigorous vibration of the hair cells in the *cochlea*, the snail shaped part of the inner ear, or with the three *semicircular canals* in the inner ear that help us maintain our balance.
   b. Sample answer: A pendulum that swings farther than another is said to have a higher *amplitude*.
   c. Sample answer: This statement about one pendulum moving faster than another is probably referring to *frequency*. Frequency is the number of times that the pendulum makes a complete back-and-forth motion each second.

2. Answer:
   a. Sample answer: The ruler is moving back and forth with a frequency of 30 times per second, producing a *low-pitched sound*. The *amplitude* of the vibration is associated with the loudness of the sound produced, whether sensed as noise level or measured in *decibels*.
   b. Sample answer: This statement is probably referring to the *frequency* of the tuning forks. Doubling the frequency of the sound produces the same note but an *octave* higher.
   c. Sample answer: This statement may be referring to *echoes* or to *resonance*. Resonance would occur when the frequency of the sound made by the person matched the natural frequency of the air column in the bucket.

3. Answer:
   a. Sample answer: *Echoes* and *resonance* may occur in a shower, enhancing the sound of the human voice.
   b. Sample answer: "Soft" and "mellow" are words that refer to the *amplitude* (or loudness) and *timbre* (or quality) of the sound. This quality may also be associated with *low-pitched sounds*.
   c. Sample answer: The need for earplugs may refer to excessive or high-*decibel noise*.

4. Answer:
   a. Sample answer: The song box, or *syrinx*, is where sounds are produced by birds.
   b. Sample answer: Disturbance of the air by a moving car creates sound that can be *reflected* from buildings and other objects. The sound that returns to the ear is an *echo*.
   c. Sample answer: Bats use *echolocation* to locate objects such as insects.

5. Answer:
   a. Sample answer: The tiny vibrating hair cells are located in the *cochlea*.
   b. Sample answer: The pegs in a guitar are wound to change the *tension* in the guitar strings and hence the *frequency* at which they vibrate.
   c. Sample answer: Knowing the *speed of sound* and the time it takes for the sound to cross the lake, the distance across the lake can be determined.

6. Answer:
   Sample answer: The *vibration* of the guitar string produces *sound* of a certain *pitch*. The pitch depends on the length of the string. By moving her fingers up the string, Cecile shortens it, causing the string to vibrate at a higher pitch.

7. Answer:
Sample answer: Bats use *echolocation* to detect sound waves of a high *frequency* that humans cannot hear.

Correction/Completion

8. Answer:
The shorter a pendulum, the *smaller* the time of the vibration and the higher the frequency of the vibration.

9. Answer:
The faster a string vibrates, the *higher* the generated sound. Or: The *greater the force with which* a string vibrates, the louder the generated sound.

10. Answer:
All of these things might cause high-pitched sounds: a very short pipe; a *short*, tight rubber band; a thin, *tight* violin string; a *small* vibrating object; and a small object dropped on the floor.

11. Answer:
The *quality* of a sound from a vibrating object results from the object vibrating in more than one way at a given time.

Short Essay

12. Answer:
Sample answer: Pluck a guitar string. Plucking the string would cause it to vibrate, which would make a sound. The younger student could listen to the guitar as the sound died away. He or she could see that as the string vibrates less and less forcefully, the sound becomes fainter and fainter. Another way to demonstrate this would be to place the sound-collecting end of a stethoscope in a bucket of water. While the student listens through the stethoscope, tap on the bucket. The sound should be clearly audible.

13. Answer:
Sample answer: Have the student place his or her hands over his or her ears. Place the handle end of a vibrating tuning fork against his or her forehead or temple. The student should be able to hear the sound clearly.

14. Answer:
Sample answer: Demonstrate sounds made by similar instruments that differ primarily in size. For example, a double bass compared with a violin, a piccolo compared with a flute, a mandolin compared with a guitar, alto and baritone saxophones, tuba and cornet, and so on.

15. Answer:
Sample answer: Let the student experiment by plucking a stringed instrument with different forces. He or she should see that the more forcefully the string is plucked, the more vigorously it vibrates, and the more vigorously it vibrates, the louder the sound it produces. Another approach would be to have him or her touch a loudspeaker playing some sort of soft sound. The volume could then be raised gradually until it was quite loud. He or she should sense a much greater vibration at a high volume than at a low volume.

16. Answer:

Sample answer: A tuba has a very low-pitched, mellow sound. On the other hand, a trumpet has a relatively high-pitched sound that may have a very piercing quality. With both instruments, the vibration is produced by the "buzzing" of the performer's lips into the mouthpiece. By tightening or loosening the lips, higher or lower tones can be produced. Also, the valves of the instruments may be operated in various combinations to direct the vibrating air into different pathways. The longer the pathway through the instrument, the lower the pitch of the tone. Louder tones are produced by forcing more air through the instrument.

17. Answer:

Sample answer: Use a sound source that emits a sharp sound once every second (clapping hands would work). Then locate a sound-reflecting structure such as a wall or cliff face. Find the shortest distance, in meters, at which the returning sound is heard at the same time as the emitted sound. Multiply that distance by 2. The answer is the speed of sound in meters per second.

Short Response

18. Answer:

Answers will vary but should stress that the light travels much faster (about 300,000,000 m/s) than the sound (a little over 300 m/s).

19. Answer:
   a. The hammer striking the metal of the bell; high frequency; small amplitude
   b. Blowing across the opening of the piccolo vibrates the air column within it; high frequency; small amplitude
   c. Falling water striking the water and riverbed below; many frequencies, from low to high; large amplitude

20. Answer:
   a. The rapid expansion of the gas inside the blank charge of the pistol as it explodes; mix of frequencies from high to low; large amplitude
   b. The vibration of the tuba player's lips, which causes the air inside the tuba to vibrate; low frequency; large amplitude
   c. The explosion of the gases inside the engine, the operation of the mechanical components inside the engine, the vibration of the exhaust pipes, and so on; mix of frequencies, from high to low; large amplitude
   d. The vibration of the metal disks that make up the cymbals; mix of frequencies, from high to low—average frequency depends on size of cymbals; large amplitude

21. Answer:
   a. long violin string
      low-pitched note
   c. thick violin string
      low-pitched note
   d. large amplitude
      loud sound
   f. long suspended iron pipe
      low-pitched sound

Graphic

22. Answer:
    a. The longer the pendulum, the lower its frequency.
    b. A pendulum with a length of about 23 cm would be a good choice because the pendulum would have a frequency of 1 vibration per second. However, some students may point out that other pendulum sizes could be used if the clock mechanism converts the motions appropriately.
    c. Yes; he worked systematically, tested one variable at a time, kept good records, and presented his data in a manner that was easy to interpret.

23. Answer:
    a. The time between strokes of the drumstick is 0.5 seconds. The interval between striking the drum (hand down) and the maximum lift of Maria's hand is 1/2 stroke, which takes 0.25 seconds. Phil is positioned so that the sound reaches him 0.25 seconds after it is produced at the source, which is when Maria's hand is raised.
    b. Maria beat the drum at a steady rate of 2 beats per second. This was their "stopwatch."
    c. See graph below. Yes; yes
    d. Phil: 85 m/0.25 s = 340 m/s; Dwayne: 165 m/0.50 s = 330 m/s; Elsie: 260m/0.75 s = 347 m/s; Vo: 335 m/1 s = 335 m/s; Leo: 500 m/1.50 s = 333 m/s; average of all observers, 337 m/s
    e. At about 417 m

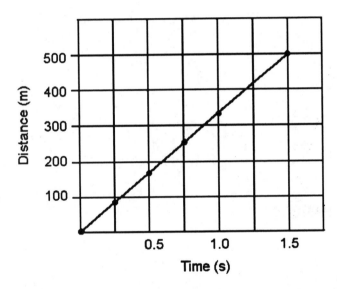

Illustrative

24. Answer:
    a. All but G and H would probably be classified as musical sounds. A, B, C, D, F and I are pure notes. G and H represent noises.
    b. A has the least amplitude and therefore could be described as the "softest" sound. B has the lowest frequency, and is therefore the lowest-pitched sound. F has the highest frequency, and is therefore the highest-pitched sound.
    c. The sound in F has a frequency four times higher than the sound in A. Therefore, if the sound represented on the oscilloscope screen in A has a frequency of 200 Hz, the sound represented in F must have a frequency of 800 Hz.

25. Answer:
    a. Which one vibrates the fastest?   a
    b. Which one vibrates the slowest?   b
    c. Which one produces the lowest sound?   b
    d. Which one produces the highest sound?   a
    e. Which one produces the loudest sound?   d

26. Answer:
    (a) The eardrum should be farther into the ear canal. (b) The
    semicircular canals are in the wrong place—they should be part of the
    inner ear, not the middle ear. (c) The stirrup, anvil, and hammer are
    in the wrong place and are mistakenly labeled as part of the inner
    ear—they should form the middle ear. (d) From outermost to innermost:
    hammer, anvil, and stirrup should be the order of the bones of the
    middle ear. (e) The nerve should be connected to the cochlea, not the
    anvil. (f) The nerve travels to the brain, not the other ear. (g) The
    Eustachian tube is connected to the throat, not the nose. (h) The
    sound waves are incorrectly drawn. They should be shown going into the
    ear, not leaving it.

27. Answer:
    The smaller tuning fork would produce sound with a shorter wavelength, not a longer
    wavelength.

28. Answer:
    a.

    b.

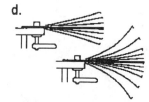

    c.

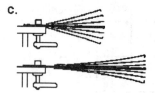

    d.

    e.

29. Answer: Sample drawings:

a.

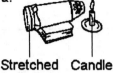

Stretched   Candle
balloon

b. Spoons

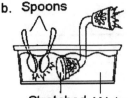

Stretched Water
balloon

c.

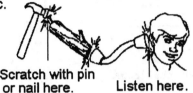

Scratch with pin
or nail here.   Listen here.

d.

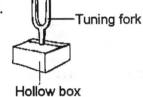

Tuning fork

Hollow box

Numerical Problem

30. Answer:
    5 s/10 claps = .5 s per clap
    Distance/.5 s = 340 m/s
    Distance = 340 m/s x 0.5 s = 170 m
    170 m ÷ 2 = 85 m

Performance Task

31. Answer:
    a. Sample hypothesis: Increasing the mass of the bob will increase the number of vibrations that the pendulum makes per second.
    b. Variables that should be held constant include the height at which the bob is released and the length of the string.
    c. Results will vary depending on pendulums used.
    d. Accept all reasonable responses.
    e. Students' conclusions should be clearly stated and well supported by data. Conclusions should either support or disprove the hypothesis.

32. Answer:
    Answers will vary but should be logical and clearly reflect what the students learned in this unit.

Chapter 19 Assessment

Word Usage

1. Use all of the following words in a sentence about energy conversion: *chemical*, *heat*, *light*, *energy*, and *firecracker*.

_____

_____

_____

Correction/Completion

2. The following sentences are incorrect or incomplete. Your challenge is to make them correct and complete.

   a. Light has mass because it is a form of matter.

_____

_____

   b. A _____ separates _____ light into a rainbow of

   colors, or a _____.

Short Response

3. Match the color or combinations of colors at left with the correct type of color of light at right.

   _____ cyan   a. primary

   _____ magenta and red   b. secondary

   _____ magenta   c. complementary

   _____ red

   _____ yellow and blue

   _____ blue

4. As solids such as steel are heated to different temperatures, they change colors. Number the following colors from coolest to hottest. Use 1 for coolest and 4 for hottest.

   a. orange   _____

   b. red   _____

   c. white   _____

   d. yellow   _____

Illustrative

5. Draw the missing prism in each experiment below. Be sure that the prism you draw is positioned correctly to match the results of the experiment.

a.

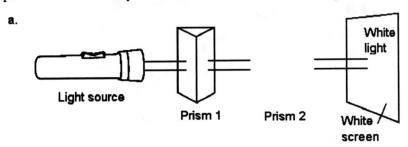

b.

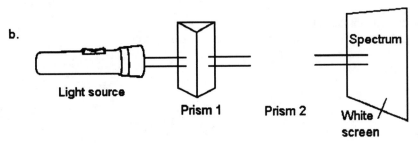

Short Response

6. You are watching a school stage production. Suddenly the stage turns blue as George moves a blue filter over the white spotlight. Jeanette remarks, "That's a pretty blue that George added to the light." Explain why Jeanette's remark is not scientifically correct.

_____

_____

_____

_____

_____

Answers to Chapter 19 Assessment

Word Usage

1. Answer:
   Sample answer: When a *firecracker* explodes, *chemical energy* is converted into *heat* energy and *light* energy.

Correction/Completion

2. Answer:
   a. Light *does not have* mass because it is a form of *energy*.
   b. A __prism__ separates __white__ light into a rainbow of colors, or a __spectrum__.

Short Response

3. Answer:
   __b__ cyan
   __c__ magenta and red
   __b__ magenta
   __a__ red
   __c__ yellow and blue
   __a__ blue

4. Answer:
   a. orange   __2__
   b. red   __1__
   c. white   __4__
   d. yellow   __3__

Illustrative

5. Answer:
   a.

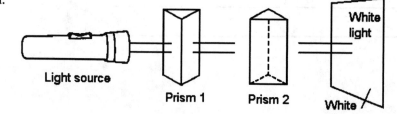

   b.

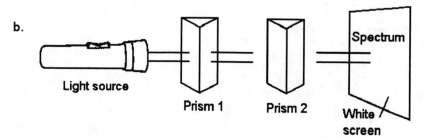

Difficulty: 3

6. Answer:

Sample answer: The blue filter does not add blue to the white light. Instead, the blue light is a result of the filter absorbing all other colors except blue.

Difficulty: 4

Chapter 20 Assessment

Word Usage

1. Write a sentence to show how the following words are related to one another: *smoke, white light, scatter*, and *blue*.

   _____

   _____

2. Which of the following terms is being described by each of the situations below: *specular reflection, diffuse reflection, scattering*?

   a. A piece of wax "lights up" when light falls on it.

   _____

   b. A flashlight shines on the walls of a room.

   _____

   c. Helen looks in the rearview mirror of her bike.

   _____

Correction/Completion

3. The following sentences are incorrect or incomplete. Your challenge is to make them correct and complete.

   a. The pebbled glass in a shower stall door is _____

   because it _____ some light but also _____ some

   of it so that objects are not clearly visible through the door.

   b. The primary colors of paint are _____, _____,

   and _____. When they are mixed together equally, the

   resulting color is _____.

   c. A building painted in dark colors is cooler than a building painted in light colors.

   _____

   _____

   _____

Illustrative

4. a. Draw and label the angle of reflection in the following illustration.

   b. What is the angle of reflection?

Incident beam

50°

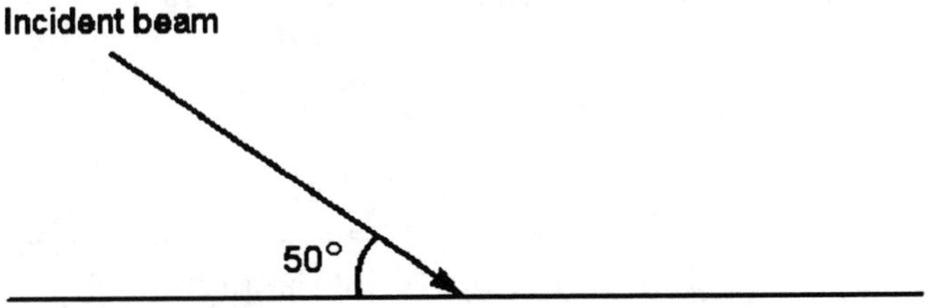

Short Response

5. While riding on a well-lighted subway train, Elvis noticed that when he stared at the window, he could see himself, and he could also see the inside of the subway tunnel.

   a. Explain why Elvis saw what he did.

   _____

   _____

   _____

   _____

   _____

   b. When the subway train stopped at a station, Elvis could see things outside the car much more clearly, but now he could barely see his reflection. Explain why.

   _____

   _____

   _____

   _____

   _____

6. Ed wanted to see if he could determine the height of the school's flagpole using his knowledge of reflected light and similar triangles. Ed learned in math class that if the corresponding angles of two triangles are congruent (they have the same measure), the triangles are *similar*. For example, the two triangles below are similar.

If two triangles are similar, the ratios of the corresponding sides are the same. In the triangles below,

$$\frac{MN}{PQ} = \frac{NO}{QR} = \frac{MO}{PR}$$

As shown in the illustration below, the flagpole is 15 m from the mirror on the ground, and Ed is 6 m from the mirror. Ed can see the top of the flagpole in the mirror. His eyes are 2 m above the ground.

a. Are angles *BCA* and *ECD* the same? Why or why not?

_____

b. Are the triangles similar? Explain.

_____

_____

_____

_____

c. How tall is the flagpole? To solve, fill in the missing values in the following ratio. Show your work.

$$\frac{AB}{BC} = \frac{DE}{CE}$$

_____

_____

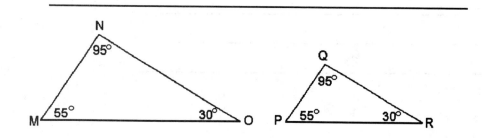

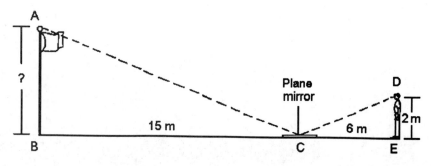

Answers to Chapter 20 Assessment

Word Usage

1. Answer: Sample answer: *Smoke* causes the *blue* part of *white light* to *scatter*.

2. Answer:
   a. Scattering
   b. Diffuse reflection
   c. Specular reflection

Correction/Completion

3. Answer:
   a. The pebbled glass in a shower stall door is <u>translucent</u> because it <u>transmits</u> some light but also <u>scatters</u> some of it so that objects are not clearly visible through the door.
   b. The primary colors of paint are <u>blue</u>, <u>yellow</u>, and <u>red</u>. When they are mixed together equally, the resulting color is <u>black</u>.
   c. A building painted in *light* colors is cooler than a building painted in *dark* colors.

Illustrative

4. Answer:
   a. See diagram below.
   b. _____50°_____

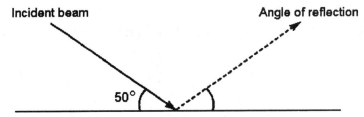

Short Response

5. Answer:
   a. Sample answer: Some of the light from inside the car was reflected by the window, so Elvis could see himself. However, some of the light from inside the car passed through the window, so Elvis could also see the subway tunnel.
   b. Sample answer: The station is well lighted, but the tunnel was very dark. Light from the station passes through the window to Elvis's eyes so he can see things in the station clearly, but this light interferes with his ability to see his own reflection.

   Difficulty: 3

Numerical Problem

6. Answer:
   a. Yes. The angle of incidence equals the angle of reflection.
   b. Yes. Angles *BCA* and *ECD* are congruent, and the angle that Ed makes with the ground (90°) is congruent to the angle that the flagpole makes with the ground (90°). Because there are 180° in a triangle, angle *BAC* must be congruent to angle *CDE*.
   c. x/15 = 2/6; 6x = 30; x = 5 m

   Difficulty: 4

Chapter 21 Assessment

Word Usage

1. a. Use all of the following terms in one or two sentences to show how they are related: *converging lens, focal point,* and *focal length.*

   _____

   _____

   _____

   b. Use all of the following terms in one or two sentences to explain the difference between a plane mirror and a projector: *object, virtual image, real image,* and *reflection.*

   _____

   _____

   _____

   _____

Short Response

2. Show how each of the following words and names would appear in a plane mirror.

   a. TOT   _____

   b. HANNAH   _____

   c. MOM   _____

   d. HOLLOW   _____

3. Name the device, either a convex mirror or converging lens, that is used

   a. to help security officers protect department stores.   _____

   b. in a camera.   _____

   c. to view the highway behind you while you ride in the passenger seat of a car.   _____

   d. to correct the vision of a farsighted person.   _____

Illustrative

4. The illustrations below show three light rays entering and exiting two lenses. Redraw the illustrations to show the correct paths taken by the light in each case.

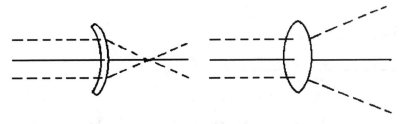

5. Some people need eyeglasses that correct for both nearsightedness and farsightedness. These eyeglasses have bifocal lenses, as shown below.

   a. Why are they called bifocal lenses?

   _____

   _____

   b. Is the top part convex or concave?

   _____

   c. Is the bottom part convex or concave?

   _____

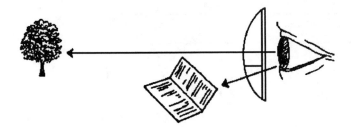

Data for Interpretation

6. The amount by which a material can bend light is given by its *index of refraction*. The index of refraction for a given material is calculated by dividing the speed of light in a vacuum (about 300,000 km/s) by the speed of light in that material.

Indices of Refraction of Common Substances

Material	Index of refraction
air	1.00
ice	1.31
water	1.33
quartz	1.46
glass	1.52
amber	1.54
ruby	1.76
diamond	2.42

a. Through which material does light travel slowest? Explain your answer.

_____

_____

b. Based on your answer to (a), what do you conclude about the relationship between speed of light through a material and the material's ability to refract light?

_____

_____

c. Through which material does light travel at about 200,000 km/s? Show your work.

_____

_____

d. The speed of light in air is not quite as fast as the speed of light in a vacuum. Why then, do you think, is the index of refraction for air 1.00?

_____

_____

_____

Answers to Chapter 21 Assessment

Word Usage

1. Answer:
   a. The *focal length* of a *converging lens* is the distance from the center of the lens to the *focal point*.
   b. Sample answer: When you look at an *object* in a plane mirror, the *reflection* you see of the *object* is a *virtual image*. When you watch a film, the projector shows a *real image* of each *object*.

Short Response

2. Answer:

   a. TOT   <u>   TOT   </u>

   b. HANNAH   <u>   HAИИAH   </u>

   c. MOM   <u>   MOM   </u>

   d. HOLLOW   <u>   WO⅃⅃OH   </u>

3. Answer:
   a. to help security guards protect department stores.   <u>Convex mirror</u>
   b. in a camera.   <u>Converging lens</u>
   c. to view the highway behind you while you ride in the passenger seat of a car.   <u>Convex mirror</u>
   d. to correct the vision of a farsighted person.   <u>Converging lens</u>

Illustrative

4. Answer:

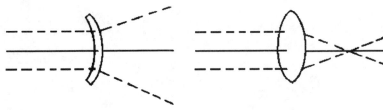

5. Answer:
   a. Because there are two focal points.
   b. Concave
   c. Convex

Difficulty: 3

6. Answer:
   a. Diamond. It has the largest index of refraction, so its speed must be the smallest of all the materials on the table.
   b. The slower light travels through the material, the greater the material's ability to refract light.
   c. 300,000 km/s ÷ 200,000 km/s = 1.5
      The material that has an index of refraction closest to this value is glass.
   d. The difference in speed is too small to be significant at two decimal places.

Difficulty: 4

Unit 7 End-of-Unit Assessment

Word Usage

1. Write a sentence to show how the following terms and phrases are related to one another: *glass, paraffin, translucent, cannot see objects clearly, transparent,* and *can see objects clearly.*

_____

_____

_____

Correction/Completion

2. Read each statement. If the statement is correct, write *correct.* If the statement is incorrect, rewrite it so that it is correct.

   a. White light is shining on a white screen in a darkened room. Both a green and red filter are placed in the path of the beam. The screen becomes black.

   _____

   _____

   _____

   b. Green paper appears green when white light falls on it because the green paper absorbs green light and reflects all other colors.

   _____

   _____

   _____

Short Response

3. Which of the following terms best describes each of the situations below: specular reflection, diffuse reflection, refraction, total internal refraction?

   a. From his canoe, Gene observed the location of an underwater jagged rock in the lake.

   _____

   b. Light is piped through curved glass fibers; dots of light appear at the end of the fibers.

   _____

   c. Marianne noticed that the reflected light was scattered when she shined the flashlight on the ground.

   _____

Short Essay

4. The primary light colors (red, green, and blue) combine to make white light, but the primary colors of paint (red, blue, and yellow) combine to make black paint. Explain.

   _____

   _____

   _____

   _____

   _____

   _____

Illustrative

5. Use the diagram below to answer the questions that follow.

   a. If you shine a beam of light (*A* in the diagram), which of the lines would represent the reflected beam?

   _____

   b. Suppose that *B* (in the diagram) is a reflected beam from some incident beam *Q*. The path of incident beam *Q* would match the path of one of the reflected beams shown. Which one?

   _____

   c. One of the reflected beams has the same path as its incident beam. Which one is it?

   _____

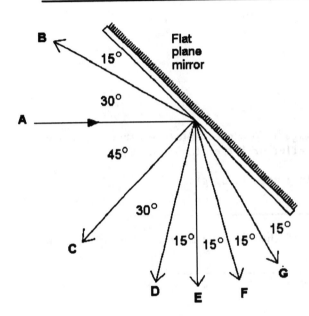

6. Examine the triangle *XYZ* shown below. If you held a plane mirror perpendicular to the page at point *Z*, how would the triangle appear in the mirror? Answer with a sketch in the space to the right of triangle *XYZ*.

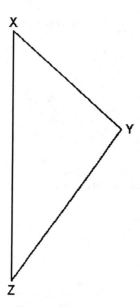

Short Response

7. In each of the following situations, an image is produced. State whether the image is real or virtual, erect or inverted, and smaller or larger than the object.

a. Mike closely examines a grasshopper with a magnifying glass.

_____

b. Mikki, the truck driver, makes good use of the two side mirrors when backing up.

_____

c. Mia takes a photograph of Claude.

_____

d. Joellen watches a film projected on a large movie screen.

_____

Data for Interpretation

8. The table below shows the distances of objects and images from a concave mirror as well as the heights of the objects and the images. Use the table to answer the questions that follow.

Situation	Object distance (cm)	Image distance (cm)	Height of object (cm)	Height of image (cm)
A	30	15	6	3
B	20	20	6	6
C	13.33	40	6	18
D	15	?	6	12

a. In situation A, how does the image distance compare with the object distance? How does the height of the image compare with the height of the object?

_____

_____

_____

b. Answer the same two questions in part (a) for situations B and C.

_____

_____

_____

c. Do you see any relationship between distances and heights? Explain.

_____

_____

_____

d. Use the relationship you discovered in part (b) to find the image distance in situation D.

_____

_____

_____

Short Essay

9. Susan is taking her younger brother, Daniel, for a walk along the beach near their home. Daniel asks why both the ocean and the sky are blue. What might Susan say to answer his question?

_____

_____

_____

_____

_____

Illustrative

10. Place one of the following symbols in each box of the diagrams below, and connect the path of the light in each. The arrow in each part indicates the direction of the light beam striking the optical device.

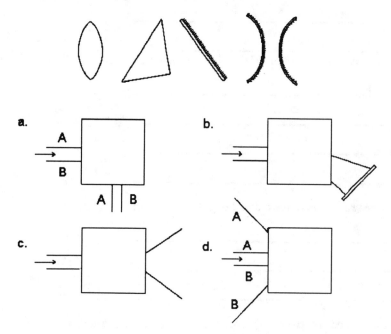

Answers to Unit 7 End-of-Unit Assessment

Word Usage

1. Answer:
   Sample answer: *Glass* is *transparent*, which means that you *can see objects clearly through it*. *Paraffin* is *translucent*, which means that you *cannot see objects clearly through it*.

Correction/Completion

2. Answer:
   a. Correct
   b. Sample answer: Green paper appears green when white light falls on it because the green paper absorbs *all colors except green, which it reflects*.

Short Response

3. Answer:
   a. Refraction
   b. Total internal refraction
   c. Diffuse reflection

Short Essay

4. Answer:
   Sample answer: The primary colors of light combine to form white light because white light is actually "built" from all of these colors. However, when you see a certain color of paint, that color is absorbing all colors except the one that you see. If all primary colors of paint are mixed, all colors are absorbed, leaving black.

Illustrative

5. Answer:
   a. Line *E*
   b. Line *G*
   c. Line *C*

6. Answer:

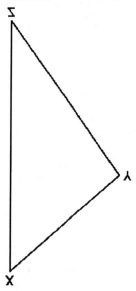

Short Response

7. Answer:
   a. The image is virtual and erect and appears larger than the object.
   b. The image is virtual and erect and appears smaller than the object.
   c. The image is real and inverted and appears smaller than the object.
   d. The image is real and inverted and appears larger than the object.

   Difficulty: 3

Data for Interpretation

8. Answer:
   a. The image distance is one-half the object distance. The height of the image is one-half the height of the object.
   b. In situation *B*, the image distance and object distance are the same; the image height and the object height are the same. In situation *C*, the image distance is three times the object distance; the image height is three times the object height.
   c. In each case, the ratio of the image distance to the object distance and the ratio of the image height to the object height are the same.
   d. The image distance is 30 cm.

   Difficulty: 4

Short Essay

9. Answer:
   Sample answer: The sky appears blue because tiny particles in the atmosphere scatter much of the blue light so that almost everywhere you look, you can see this color. The ocean has a mirrorlike surface that reflects the blue light of the atmosphere. Thus, the ocean appears blue.

   Difficulty: 4

Illustrative

10. Answer:

a.

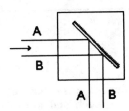

b.

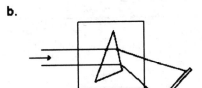

c.

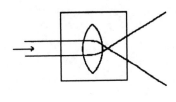

d.

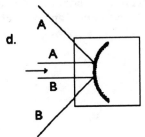

Difficulty: 4

Unit 7 Activity Assessment

Activity Assessment
1. It's All Done With Mirrors
Teacher's Notes
Overview
Students use their knowledge of incident and reflected beams to design a system for viewing the reflection of light from a flashlight. They produce a sketch and a table of values for the different angles of incidence and reflection used.

Materials
(per activity station)
- a flashlight
- 2 shoe boxes
- 5-8 plane mirrors
- masking tape
- modeling clay
- a metric ruler
- a protractor

Preparation
Prior to the assessment, equip student activity stations with the materials needed for each experiment. Prepare the shoe boxes as shown below.

Time Required
Each student should have 30 minutes at the activity station and 10 minutes to complete the Data Sheet.

Performance
At the end of the assessment, students should turn in the following:
- a completed Data Sheet

Evaluation
The following is a recommended breakdown for evaluation of this Activity Assessment:
- 20% appropriate use of equipment and materials
- 40% ability to successfully make use of the angles of incidence and reflection in a plane mirror
- 20% clarity and accuracy of sketch
- 20% precise and accurate recording of data

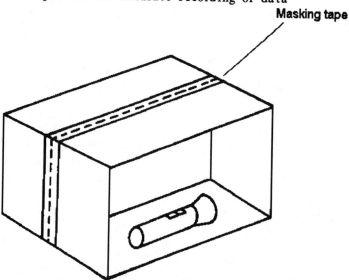

Masking tape

2. It's All Done With Mirrors

Using mirrors, magicians can make you see things that aren't really in front of you. Now it's your turn to do magic—and all you need are plane mirrors and some careful concentration! You'll have to calculate the angles of the incident beams and reflected beams in order to move an image out of one shoe box and into another. And it really can be done with mirrors.

Before You Begin . . .

As you work through the tasks, keep in mind that your teacher will be observing the following:
• how you use the materials and equipment
• how thoughtfully and logically you arrange your mirrors
• your ability to accurately record data and draw your sketch

Work Your Magic!

Task 1 Place a flashlight inside one of the shoe boxes. The light should not shine out of the shoe box when you turn the flashlight on.

Task 2 Using the materials at your activity station, design an arrangement of plane mirrors that redirects the beam of the flashlight. You should be able to turn the flashlight on in one shoe box and view its reflection in a mirror placed in the corner of the other shoe box. You may not need to use all of the plane mirrors provided.

Task 3 On your Data Sheet, sketch a diagram of your setup, including the shoe boxes, the arrangement of the mirrors, and the path of the light beam from the point where it leaves the flashlight to the point where it reaches your eye. Label each mirror (Mirror 1, Mirror 2, etc.).

Task 4 Using the table on your Data Sheet, record the angles of incidence and reflection for each mirror you used in your design.

3.
Sketch:

Values for Angles of Incidence and Reflection

Mirror	Angle of incidence	Angle of reflection
1		
2		
3		
4		
5		
6		
7		
8		

Answers to Unit 7 Activity Assessment

Activity Assessment

1. Answer: Not applicable (teacher's notes)

2. Answer: Not applicable (student's notes)

3. Answer:
   Sketches will vary but should clearly depict and label the angles of incidence and reflection.

Unit 7 SourceBook Assessment

True/False

1. James Maxwell's discovery that light travels by electromagnetic waves was somewhat of an accident.

   a. true   b. false

2. The light that we see is part of the electromagnetic spectrum.

   a. true   b. false

Multiple Choice

3. In 1905, Albert Einstein discovered that light acted as if it were made of particles. These particles of light are called

   a. electrons.   b. photons.   c. protons.   d. quarks.

4. The presently accepted theory for the behavior of light is the

   a. particle theory.   b. wave theory.   c. particle-wave theory.

True/False

5. Visible light makes up over 50 percent of the electromagnetic spectrum.

   a. true   b. false

Multiple Choice

6. The properties of an electromagnetic wave depend on the wave's

   a. amplitude.   b. intensity.   c. frequency.   d. speed.

7. Waves that carry electrical power through transmission lines and cause static on car radios are called

   a. power waves.   b. radio waves.   c. gamma rays.   d. microwaves.

8. Which type of radio waves are reflected by the ionosphere and can be received far away from the broadcast station?

   a. AM   b. FM   c. radar waves   d. microwaves

9. The lowest frequency of visible light appears

   a. blue.   b. orange.   c. green.   d. red.   e. violet.

10. The electromagnetic rays used in cancer treatment are called

    a. X rays.
    b. infrared rays.
    c. gamma rays.
    d. ultraviolet rays.

11. Light travels at a speed of 3.0×10^8 m/s in

    a. a vacuum.   b. water.   c. air.   d. a solid.

12. The military uses equipment that enables objects to be seen in complete darkness. What electromagnetic waves make this possible?

    a. visible light
    b. X rays
    c. infrared rays
    d. ultraviolet rays

13. Laser light is the result of numerous light waves of the same frequency that are in step. This is accomplished by

    a. polarization.
    b. holography.
    c. stimulated emission.
    d. electron beams.

14. Which of the following is holography *not* used for?

    a. transmitting phone conversations
    b. studying bacteria
    c. increasing the security of bank cards
    d. studying pressure

15. Total internal reflection is used in

    a. holography.
    b. stimulated emission.
    c. fiber optics.
    d. polarization.

16. A device that carries light into otherwise unreachable places is called a(n)

    a. hologram.
    b. refractor tube.
    c. object beam.
    d. optic fiber.

17. Which of the following does *not* suggest the particle theory of light?

    a. Light passing through a small opening is diffracted.
    b. A metal emits electrons when high frequency light hits it.
    c. Light travels in straight lines.
    d. None of the above

Short Response

18. Name one use for each of the electromagnetic waves listed below.

    a. radio _____

    b. infrared _____

    c. X ray _____

    d. gamma ray _____

19. Place the following waves in order of increasing wavelength: *ultraviolet light, microwaves, gamma rays, radio waves,* and *visible light.*

    _____

    _____

20. Arrange the following colors of light in order of increasing frequency: *green, red, violet, orange, blue,* and *yellow.*

    _____

    _____

21. Arrange the following materials in order of increasing index of refraction: *water, diamond, air,* and *glass.*

    _____

    _____

22. Explain the experiment performed by Thomas Young in 1801 that caused the wave theory of light to be accepted over Newton's particle theory.

    _____

    _____

    _____

    _____

    _____

23. Explain why AM radio broadcasts can be heard at greater distances from the transmitting radio station than FM radio broadcasts.

    _____

    _____

    _____

    _____

24. Ultraviolet light can cause sunburn and, with extended exposure, skin cancer. Name two beneficial effects of ultraviolet light.

_____

_____

_____

_____

25. How do polarized sunglasses decrease glare?

_____

_____

_____

_____

Answers to Unit 7 SourceBook Assessment

True/False

1. Answer: a. true

2. Answer: a. true

Multiple Choice

3. Answer: b. photons.

4. Answer: c. particle-wave theory.

True/False

5. Answer: b. false

Multiple Choice

6. Answer: c. frequency.

7. Answer: a. power waves.

8. Answer: a. AM

9. Answer: d. red.

10. Answer: c. gamma rays.

11. Answer: a. a vacuum.

12. Answer: c. infrared rays

13. Answer: c. stimulated emission.

14. Answer: a. transmitting phone conversations

15. Answer: c. fiber optics.

16. Answer: d. optic fiber.

17. Answer: a. Light passing through a small opening is diffracted.

Short Response

18. Answer:
    Sample answers:
    a. radio   To transmit TV and radio signals
    b. infrared   To "see" objects in total darkness
    c. X ray   To see inside the human body
    d. gamma ray   To treat cancer

19. Answer: Gamma rays, ultraviolet light, visible light, microwaves, radio waves

20. Answer: Red, orange, yellow, green, blue, violet

21. Answer: Air, water, glass, diamond

22. Answer:
    Young made two slits in a card. He placed a light source on one side and a screen on
    the other. When the light was turned on, a pattern of bright and dark bands was seen on
    the screen. This pattern could only result from two waves interfering with each other.
    This proved that light has properties of waves, and it caused the wave theory of light
    to be accepted over Newton's particle theory.

23. Answer:
    Sample answer: AM radio waves have longer wavelengths than FM radio waves and are
    reflected by the ionosphere. The reflected AM waves can be received past the curve of
    the Earth and therefore can be heard at great distances from their source. The shorter
    FM radio waves are not reflected but pass through the ionosphere.

24. Answer:
    Sample answer: Skin cells produce vitamin D when exposed to small amounts of
    ultraviolet light. UV light is also used to kill germs.

25. Answer:
    Light travels in transverse waves of differing angles. Polarized sunglasses use filters
    to stop all but the vertical waves. Since glare consists largely of horizontal light
    waves, the glasses decrease glare.

Unit 7 Extra Assessment Items

Word Usage

1. For each group of terms, write one or two sentences to show how the terms are related to one another.

   a. white light, prism, spectrum

   _____

   _____

   _____

   b. object, image, behind, in front of, plane mirror, distance

   _____

   _____

   _____

   c. plane mirrors, convex mirrors, concave mirrors, real images, virtual images

   _____

   _____

   _____

2. Rewrite the paragraph below by substituting the following terms for words or groups of words in the paragraph: *diffuse, transmitted, absorbed, reflected,* and *specular reflection.*

   When light energy falls on an object, some of the energy may stay with the object, causing the object to heat up. Some of the energy may bounce off the object, and some of the light energy may pass through the object. If the surface is very smooth, then reflection of the light in only one direction occurs. However, in the case of other surfaces, the light bounces off in many different directions.

   _____

   _____

   _____

   _____

   _____

   _____

   _____

Correction/Completion

3. Some of the following statements are correct, and some are not. Read each statement. If the statement is correct, write "correct." If the statement is incorrect, rewrite it so that it is correct.

   a. When a beam of light hits a convex mirror, the light is reflected to a certain point.

   _____

   _____

   _____

   b. When light energy shines on a lens or a piece of glass, all of the energy goes through it.

   _____

   _____

   _____

4. Some of the following statements are correct, and some are not. Read each statement. If the statement is correct, write "correct." If the statement is incorrect, rewrite it so that it is correct.

   a. A convex mirror has a surface that is "bulged out," and a concave mirror has a surface that is "caved in" in relation to the light hitting it.

   _____

   _____

   b. A concave mirror produces only real images of objects, while a convex mirror produces only virtual images.

   _____

   _____

5. Some of the following statements are correct, and some are not. Read each statement. If the statement is correct, write "correct." If the statement is incorrect, rewrite it so that it is correct.

a. If a real image is smaller than the object, the image is closer to the optical device than the object is.

_____

_____

b. Light travels faster in water than it travels in air.

_____

_____

c. Total internal reflection occurs when light coming from a certain angle strikes a border between two materials in which the speed of light is greater in the first material than in the second.

_____

_____

_____

_____

_____

Short Response

6. Indicate the type of mirror (plane, concave, or convex) that you could use to obtain the following images:

a. An enlarged image of your face

_____

b. A right-side-up image of your face

_____

c. An inverted image of your face

_____

7. Describe one way that you could obtain the images listed below. State the relative positions of your face, the image, and the mirror, and describe the size of your face and the image.

a. Enlarged image of your face

_____

_____

_____

_____

_____

_____

b. Right-side-up image of your face

_____

_____

_____

_____

_____

c. Upside-down image of your face

_____

_____

_____

_____

_____

_____

8. Predict the color(s) that you would observe in the following situation:

You look at the cover of your *SciencePlus* textbook in a photography room illuminated by red light only.

_____

_____

_____

_____

9. Predict the color(s) that you would observe in the following situation:

You view the flag of the United States under a deep blue light.

_____

_____

_____

_____

10. Predict the color(s) that you would observe in the following situation:

You look through deep green sunglasses at a white sweater with green and red stripes on the sleeves.

_____

_____

_____

_____

11. Predict the color(s) that you would observe in the following situation:

In a room with white walls, you cover an ordinary incandescent light bulb with a piece of transparent orange plastic and then, on top of that, a purple piece of plastic.

_____

_____

_____

_____

12. Predict the color(s) that you would observe in the following situation:

In a darkened room, you shine a beam of white light through a blue-green filter, letting it fall on a red cardboard screen.

_____

_____

_____

_____

13. Predict the color(s) that you would observe in the following situation:

In a darkened room, white light is shone through a triangular glass prism. A green filter is placed in the path of the emerging light. The resulting beam then reaches a white screen.

_____

_____

_____

_____

14. Predict the color(s) that you would observe in the following situation:

You look through deep blue sunglasses at a rainbow in the sky.

_____

_____

_____

_____

15. Traveling through the desert on a camel, Meredith believes she sees a shimmering pool of water. Circle the term below that would be used to describe this situation.

specular reflection, diffuse reflection, scattering, refraction, total internal reflection

16. Name the device (concave mirror, prism, or converging lens) that is used for each of the following:

a. making a photograph of your best friend

_____

b. concentrating the beam of a searchlight

_____

c. producing a spectrum

_____

Graphic

17. a. An object moves away from a plane mirror. Which graph might show the height of the image?

_____

   b. Which graph might show the height of the image if **a convex mirror were used**?

_____

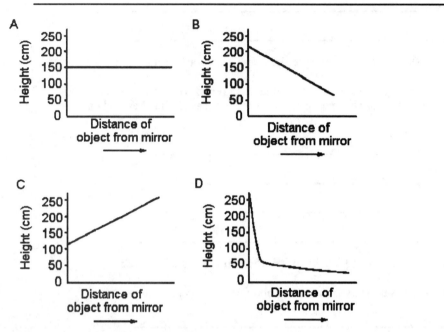

18. Which of these graphs accurately represents the relationship between the angles made by the incident and reflected beams with a plane mirror? The horizontal axis represents the angle of the reflected beam with the mirror; the vertical axis represents the angle of the incident beam with the mirror.

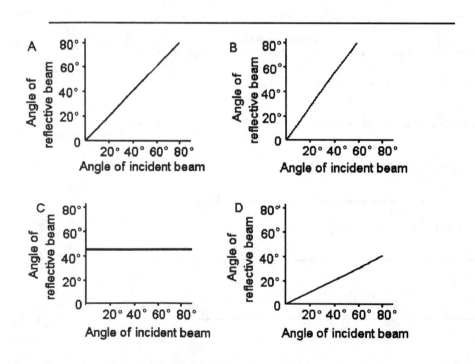

19. The illustration below shows an energy change involving light. Describe the energy change shown.

_____

_____

_____

_____

20. The illustration below shows an energy change involving light. Describe the energy change shown.

_____

_____

_____

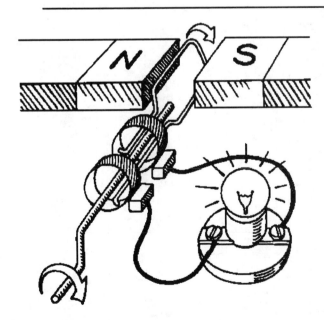

21. The illustration below shows an energy change involving light. Describe the energy change shown.

_____

_____

_____

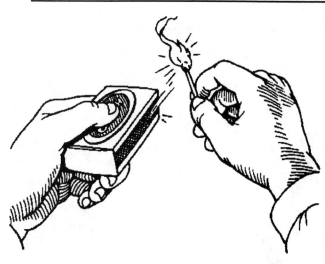

22. The illustration below shows an energy change involving light. Describe the energy change shown.

_____

_____

_____

_____

_____

23. The illustration below shows an energy change involving light. Describe the energy change shown.

_____

_____

_____

_____

_____

24. In the diagram, a beam of light hits a flat mirror and is reflected.

What is the measurement of angle A?

_____

What is the measurement of angle B?

_____

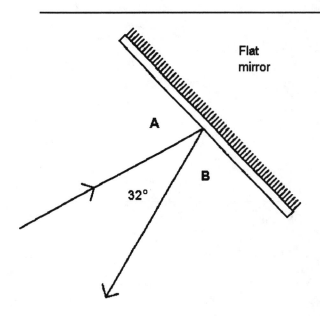

Flat mirror

A

B

32°

25. Use the diagram to answer the questions that follow.

   a. Which of these images are virtual? real?

   _____

   b. Which of these images can be projected onto a screen?

   _____

   c. Where would you place your eye to view the virtual image(s)?

   _____

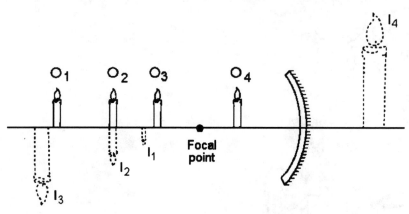

26. Complete the path of the light after hitting each of these surfaces. Some figures may have more than one answer.

   a.

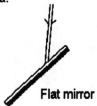

Flat mirror

   b.

Water

   c.

Air

Water

   d.

Glass

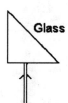

   e.

Light beams

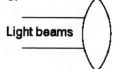

27. Look carefully at the diagram. What's wrong with it?
    Correct and complete the diagram.

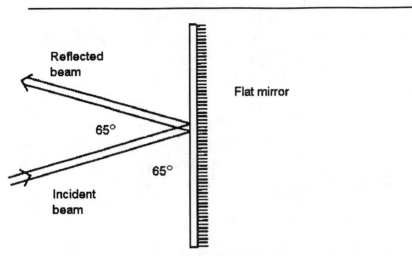

28. The diagrams show situations in which an image is produced on a screen. Is each one possible? If not, redraw the correct image.

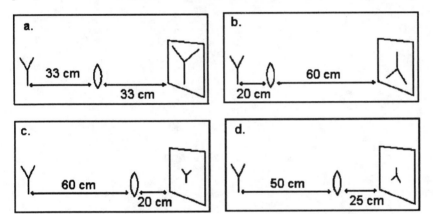

29. Draw the device (mirror type, prism, lens, etc.) that would be used for each of these situations.

    a. seeing yourself life size

    b. clearly seeing a freckle on your nose

30. Draw the device (mirror type, prism, lens, etc.) that would be used for each of these situations.

    a. taking a photograph of your best friend

    b. reading very small writing

31. Draw the device (mirror type, prism, lens, etc.) that would be used for each of these situations.

    a. a searchlight

    b. a motion picture projector

32. Draw the device (mirror type, prism, lens, etc.) that would be used for each of these situations.

   a. viewing the highway behind you when driving a car

   b. producing a spectrum

33. Draw an arrangement of two mirrors that would enable you to see a person in the next room.

Numerical Problem

34. Light travels through empty space at a speed of approximately 300,000 km/s. If a spacecraft is located 148,800,000 km from Earth, how long would it take for the light beam to travel from Earth to the spacecraft and back to Earth again? Show your work.

_____

Performance Task

35. **Materials:** double-convex lens, plain white paper or card, candle, meterstick

    Task: Use a double-convex lens to form on a plain white paper or card an image of the candle placed at a distance from the lens. Depict the object, lens, and image in a sketch.

    a. How does the image compare with the real object?

    _____

    _____

    b. Now place the candle closer to the lens and sketch what you observe.

    _____

    _____

    c. How does varying the position of the candle affect the size of the image?

    _____

    _____

    d. At what distance from the lens does the real image of the candle disappear? Why?

    _____

    _____

    _____

Answers to Unit 7 Extra Assessment Items

Word Usage

1. Answer:
   a. Sample answer: *White light* passing through a *prism* is separated into the colors of the *spectrum*.
   b. Sample answer: The *image* formed by an *object* that is a certain *distance in front of* a *plane mirror* appears to be the same distance *behind* the mirror.
   c. Sample answer: *Plane mirrors* and *convex mirrors* produce *virtual images*. *Concave mirrors* produce both *virtual images* and *real images*.

2. Answer:
   Sample answer: When light energy falls on an object, some of the energy is *absorbed*, causing the object to heat up. Some of the energy may be *reflected*, and some of the light energy may be *transmitted* through the object. If the surface is very smooth, then *specular reflection* occurs. However, in the case of other surfaces, the light that is reflected is a *diffuse* reflection.

Correction/Completion

3. Answer:
   a. Sample answer: When a beam of light hits a *concave* mirror, the light is reflected to a certain point.
   b. Sample answer: When light energy shines on a lens or a piece of glass, *most* of the energy goes through it.

4. Answer:
   a. Correct
   b. A concave mirror produces *both real and virtual* images of objects, while a convex mirror produces only virtual images.

5. Answer:
   a. Correct
   b. Light travels *slower* in water than it travels in air.
   c. Total internal reflection occurs when light coming from a certain angle strikes a border between two materials in which the speed of light is *slower* in the first material than in the second.

Short Response

6. Answer:
   a. Concave
   b. Plane, convex, or concave
   c. Concave

7. Answer:
   a. If your face is very close to a concave mirror, an enlarged, right-side-up image of your face will appear to be behind the mirror.
   b. Consider any of the following arrangements acceptable: (1) A plane mirror will produce a right-side-up image of your face that appears to be the same size and the same distance behind the mirror that you are in front of it. (2) A concave mirror can also produce a right-side-up image. The image appears larger and farther away than the object. (3) Also, the image of your face in a convex mirror will be right-side-up but will appear smaller and behind the mirror.
   c. If your face is far enough away from it, a concave mirror will produce a small, upside-down image of your face that appears to be located between your face and the mirror.

8. Answer: White and red objects will appear red. All other colors will appear black.

9. Answer:
   White areas in each flag will appear blue as will the blue portion of the flag. The red areas will appear black.

10. Answer: The white and green areas will appear green. The red stripes will appear black.

11. Answer:
   After passing through the orange plastic, only orange light would emerge. This would then be absorbed by the purple filter, which permits only the passage of purple light. Therefore, very little or no light would emerge, and everything would appear black.

12. Answer:
   Because only blue-green light will pass through the filter, the red cardboard screen will appear black.

13. Answer:
   The prism will break the white light into the full spectrum of colors. The green filter, however, will permit only the passage of the green part of the spectrum. Therefore, only the green part of the spectrum will appear on the white screen.

14. Answer: Only the blue part of the rainbow will be visible through the deep blue sunglasses.

15. Answer: refraction

16. Answer:
   a. Converging lens
   b. Concave mirror
   c. Prism

Graphic

17. Answer:
   a. *B*
   b. *D*

18. Answer: *A*

Illustrative

19. Answer:
   Sample answer: The chemical energy of a battery is converted into the electrical energy of a circuit. This electrical energy is then converted into both light and heat energy by the bulb in the circuit.

20. Answer:
   Sample answer: Mechanical energy is first converted to electrical energy. Electrical energy flowing through the wire is converted to heat and light by the bulb in the circuit.

21. Answer:
The answer is acceptable if it mentions the conversion of chemical energy or chemical potential energy to light and heat.

22. Answer:
The key idea is the conversion of light energy to chemical energy, which is stored as a simple sugar or starch within the leaves of the plant. This conversion is called photosynthesis. Any answer that represents this conversion is acceptable. If the process is described in detail, give the answer a bonus.

23. Answer:
Sample answer: Light energy is focused by the lens to create heat, which in turn initiates burning. In the process of burning, chemical energy is converted to light and heat. An answer that describes either the conversion of light to heat or chemical energy to light is acceptable. If two or more of these energy changes are mentioned, give the answer a bonus.

24. Answer:
74°
74°

25. Answer:
a. I_4 is a virtual image; the others are real.
b. I_1, I_2, and I_3 can be projected onto a screen.
c. Between the focal point and the mirror

26. Answer: Sample answers:

a.

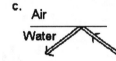

Flat mirror

b.

Water

c. Air
Water

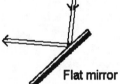

d.

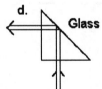

Glass

e.

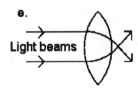

Light beams

27. Answer: The angle between the incident beam and the reflected beam is wrong.

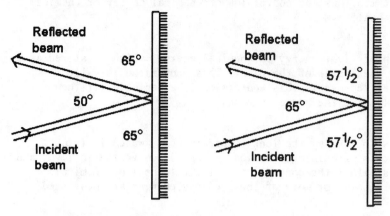

28. Answer:
   a. No; image should be inverted and same size.
   b. No; image should be inverted and enlarged. The virtual image should be three times the object's size.
   c. No; image should be inverted, one-third object size.
   d. Yes; image is inverted and about one-half object size.

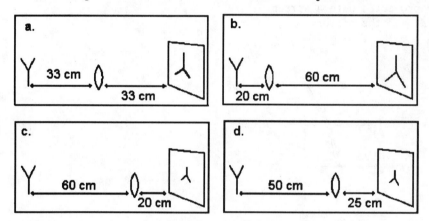

29. Answer:
   a. Students should draw a plane mirror.
   b. Students should draw a concave mirror.

30. Answer:
   a. Students should draw a converging lens
   b. Students should draw a converging lens.

31. Answer:
   a. Students should draw a concave mirror.
   b. Students should draw a converging lens.

32. Answer:
   a. Students should draw a plane or convex mirror.
   b. Students should draw a prism.

33. Answer: Sample answer:

Sample Mirror Arrangement

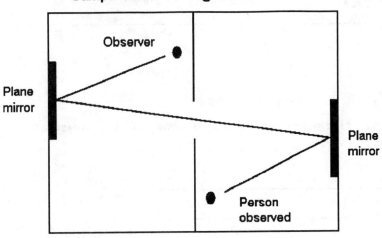

Numerical Problem

34. Answer: 2 x 148,800,000 km ÷ 300,000 km/s = 992 s, or 16 min. 32 sec.

Performance Task

35. Answer:
   a. The image is smaller and upside down.
   b. The image is inverted and larger.
   c. Moving the lens closer to the candle increases the size of the image.
   d. At the focal point. This is because the rays are parallel and do not converge to form an image.

Chapter 22 Assessment

Word Usage

1. For each group of words below, write a sentence or two to show how they are related.

   a. generation, family, traits

   _____

   _____

   _____

   b. daughter cells, mitosis, parent cell

   _____

   _____

   _____

   c. nucleus, chromosomes, genes

   _____

   _____

   _____

Correction/Completion

2. The following sentences are incorrect or incomplete. Rewrite the sentences to make them correct and complete.

   a. Cross-pollination is a form of asexual reproduction that can be used to produce seeds.

   _____

   _____

   _____

   b. Identical twins result when two eggs are fertilized by two sperm cells.

   _____

   _____

   _____

   c. Meiosis is the process that results in daughter cells with twice as many chromosomes as the parent cell.

   _____

   _____

Short Response

3. The list below includes the four essential properties of life and a definition of each property. Write the letter of the appropriate property next to each example in the space provided.

   a. Self-preservation: _____ If a starfish loses one of its legs, it can grow a new one to replace it.

   b. Self-regulation: _____ While exercising, a person's heart beats faster.

   c. Self-organization: _____ Sunflowers die at the end of the growing season, but the next year there are more sunflowers.

   d. Self-reproduction: _____ A mountain lion chases and kills a deer.

Illustrative

4. Use the family tree shown below to answer the questions that follow.

   a. Name all of Edward's uncles.

      _____

   b. Which people in the tree are Cynthia's ancestors?

      _____

   c. Name one male from each generation.

      _____

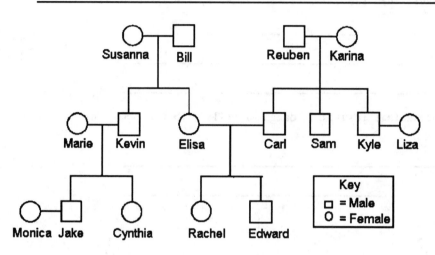

Short Essay

5. Why do you think this chapter referred to the navel as a "permanent reminder of our once parasitic mode of living"?

   _____

   _____

   _____

   _____

   _____

   _____

   _____

   _____

   _____

   _____

Answers to Chapter 22 Assessment

Word Usage

1. Answer:
   a. Sample answer: In a *family*, *traits* such as colorblindness can be passed from one *generation* to the next.
   b. Sample answer: In *mitosis*, a *parent cell* divides into two *daughter cells*.
   c. Sample answer: The *nucleus* of a cell contains *chromosomes*, which are made up of *genes*.

Correction/Completion

2. Answer:
   a. Cross-pollination is a form of *sexual* reproduction that can be used to produce seeds.
   b. Identical twins result when *one egg is fertilized* by *one sperm cell, and the resulting mass of cells separates into two halves.*
   c. Meiosis is the process that results in daughter cells with *half* as many chromosomes as the parent cell.

Short Response

3. Answer:
   <u>c</u> If a starfish loses one of its legs, it can grow a new one to replace it.
   <u>b</u> While exercising, a person's heart beats faster.
   <u>d</u> Sunflowers die at the end of the growing season, but the next year there are more sunflowers.
   <u>a</u> A mountain lion chases and kills a deer.

Illustrative

4. Answer:
   a. Kyle, Sam, and Kevin
   b. Maria, Kevin, Susanna, and Bill
   c. 1: Bill or Reuben; 2: Kevin, Carl, Sam, or Kyle; 3: Jake or Edward

   Difficulty: 3

Short Essay

5. Answer:
   Sample answer: The fetus gets all of its nutrition from the mother's body, much as a parasite gets its nutrition from the host. The nutrients are delivered to the fetus through the placenta and umbilical cord. After birth, the umbilical cord sloughs off, leaving a permanent scar—the navel.

   Difficulty: 4

Chapter 23 Assessment

Correction/Completion

1. The following sentences are incorrect or incomplete. Rewrite the sentences to make them correct and complete.

   a. Early geneticists found that genes could be colored with dye that made them easier to observe.

   _____

   _____

   _____

   b. The structure of DNA is much like a stack of coins, with each coin consisting of two of four possible chemicals.

   _____

   _____

   _____

Short Response

2. You have read that the DNA in cells can be compared with the blueprints of a factory. Match the terms at the left with those at the right to complete the comparison.

   a. Main office of factory   _____ DNA

   b. Blueprints for building new factories   _____ gene

   c. Sudden changes to the blueprints   _____ nucleus of a cell

   d. A single item from the blueprints   _____ protein

   e. A product produced by the factory   _____ mutations

Numerical Problem

3. A section of the DNA spiral ladder with 2000 rungs might make up one gene. Each of your cells contains about 100,000 genes. About how many rungs would each cell contain? Show your work.

   _____

   _____

4. One hundred babies were born at Smith General Hospital last month. There were five sets of twins and two sets of triplets born; the rest were single births. How many single births were there? Show your work.

_____

_____

Short Essay

5. Ted Miller has been accused of a serious crime. There were no witnesses, but forensic scientists matched Ted's DNA "fingerprint" to that of blood samples found at the scene of the crime. Detective Jackson announced, "Even without a witness, there can be no doubt that we have the right person." Would the detective be as confident if he knew that Ted Miller had an identical twin brother? a fraternal twin brother? Explain why or why not.

_____

_____

_____

_____

_____

_____

_____

_____

Graphic

6. Carefully examine the graph that follows, and then answer the questions that follow.

a. If neither parent is diabetic, what is the percentage of offspring with diabetes at the age of 50?

_____

b. If both parents are diabetic, what is the percentage of offspring with diabetes at the age of 50?

_____

c. Do the data in (a) and (b) suggest that diabetes is influenced by genetic factors? Why or why not?

_____

_____

_____

d. As the offspring of parents with diabetes grow older, what happens to their chances of becoming diabetic?

_____

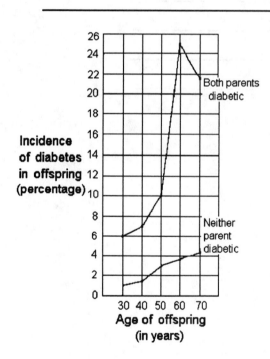

Answers to Chapter 23 Assessment

Correction/Completion

1. Answer:
   a. Early geneticists found that *chromosomes* could be colored with dye that made them easier to observe.
   b. The structure of DNA is much like a *coiled ladder or spiral staircase* with each *rung or step* consisting of two of four possible chemicals.

Short Response

2. Answer:
   <u>b</u> DNA
   <u>d</u> gene
   <u>a</u> nucleus of a cell
   <u>e</u> protein
   <u>c</u> mutations

Numerical Problem

3. Answer: 100,000 x 2000 = 200,000,000 (200 million) rungs

4. Answer: 100 - (5 x 2) - (2 x 3) = 84 single births

Short Essay

5. Answer:
   Sample answer: If Ted Miller had a fraternal twin brother, the detective would still be confident about the DNA match. Fraternal twins are produced by two different sperm cells and two different egg cells and therefore would have different genetic makeups. However, if Ted Miller had an identical twin brother, their genetic makeup would be almost identical because they both were formed from the same sperm cell and egg cell. Therefore, the detective would not be as confident.

   Difficulty: 3

Graphic

6. Answer:
   a. 3 percent
   b. 10 percent
   c. Yes; the data show that a greater percentage of people develop diabetes if they have diabetic parents.
   d. Chances increase with age until about age 60.

   Difficulty: 4

Unit 8 End-of-Unit Assessment

Word Usage

1. Colin said, "In an old family photograph album, I noticed that some of my father's ancestors had cleft chins, just like mine. But my brother Brad doesn't have one. Why is that?"

    Explain to Colin how two brothers can inherit different sets of features. Use the following words in your explanation: *dominant, gene, inherited,* and *recessive.*

    _____

    _____

    _____

    _____

    _____

2. Use the following words in one or more sentences to show how they are related.

    a. mitosis, meiosis, chromosomes, nucleus

    _____

    _____

    b. chromosomes, linear programming, genetic information

    _____

    _____

    c. code, sequence, chemicals, DNA

    _____

    _____

Correction/Completion

3. Complete the following statements with *early, midway,* or *late* to show approximately when the events of your development occurred.

   a. The palms of your hand could respond to a light touch

   _____ in pregnancy.

   b. Your sex organs developed _____ in pregnancy.

   c. Your heart began to beat _____ in pregnancy.

   d. You practiced breathing _____ in pregnancy.

   e. Your fingerprints formed _____ in pregnancy.

Illustrative

4. a. Place the following labels on the diagram: *chromosomes, nucleus,* and *cell membrane.*

   b. What process does the diagram illustrate?

   _____

   c. What molecule controls this process?

   _____

5. The following diagram shows what happens when sex cells form during meiosis. Draw in the correct number of chromosomes at each step in the cell's division.

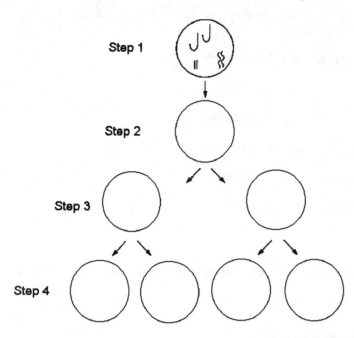

Step 1

Step 2

Step 3

Step 4

Short Response

6. Humans have body cells and sex cells. Show that you recognize the distinguishing features of these cells by writing "body cells" or "sex cells" in the space beside each description below.

a. There are only two kinds. _____

b. They contain 23 pairs of chromosomes. _____

c. They unite to form the next generation. _____

d. They contain 23 chromosomes. _____

e. There are many different kinds. _____

Illustrative

7. When Mendel crossed tall, hybrid pea plants, he noticed a pattern among the offspring. Again and again, three-fourths of the offspring showed the dominant trait (tallness) and one-fourth showed the recessive trait (shortness). Draw a Punnett square to explain this observation. Use T for the dominant factor and t for the recessive factor.

Data for Interpretation

8. Use the data in the table below to answer the questions that follow.

Timing of mother's exposure to rubella	Risk of defects in offspring
weeks 1 - 8	85 percent
weeks 9 - 12	52 percent
weeks 13 - 20	16 percent

a. What happens to the risk of defects in offspring as pregnancy advances?

_____

_____

b. When is the risk greatest?

_____

_____

_____

_____

Why do you think the risk of defects is so high during this time?

_____

_____

_____

_____

Short Response

9. Name an environmental factor that might influence a developing embryo or fetus, and describe the factor's influence.

_____

_____

_____

_____

_____

_____

_____

_____

Illustrative

10. Complete the following diagrams by filling in the correct number of chromosomes and cells that are present during each process.

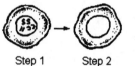

Step 1   Step 2   Step 3

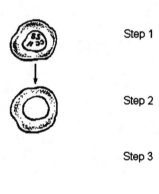

Step 1

Step 2

Step 3

Step 4

Short Essay

11. Write a paragraph about how a flowering cactus illustrates the four essential properties of life: *self-preservation, self-regulation, self-organization,* and *self-reproduction.*

---

12. Why are there many noticeable differences among a group of people but few noticeable differences among a group of bacteria of the same species? What does the method of reproduction have to do with this?

Answers to Unit 8 End-of-Unit Assessment

Word Usage

1. Answer:
   Sample answer: Many characteristics, such as the presence of a cleft chin, are determined by *genes* that are *inherited* from ancestors. Each parent contributes one gene, which can be *dominant* or *recessive*, to their offspring. Brad does not have a cleft chin because he inherited two recessive genes.

2. Answer:
   a. Sample answer: *Mitosis* and *meiosis* are two ways in which copies of the *chromosomes* in the *nucleus* are distributed to new cells.
   b. Sample answer: *Genetic information* contained in *chromosomes* could be considered a form of *linear programming*.
   c. Sample answer: The *chemicals* in *DNA* are arranged in a *sequence* that forms a *code*.

Correction/Completion

3. Answer:
   a. The palms of your hand could respond to a light touch <u>late</u> in pregnancy.
   b. Your sex organs developed <u>early</u> in pregnancy.
   c. Your heart began to beat <u>early</u> in pregnancy.
   d. You practiced breathing <u>late</u> in pregnancy.
   e. Your fingerprints formed <u>midway</u> in pregnancy.

Illustrative

4. Answer:
   a. See diagram below.
   b. Mitosis
   c. DNA

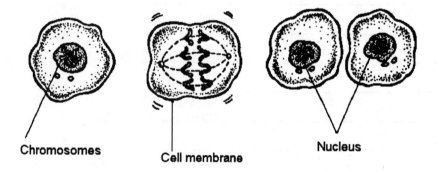

Chromosomes

Cell membrane

Nucleus

5. Answer:

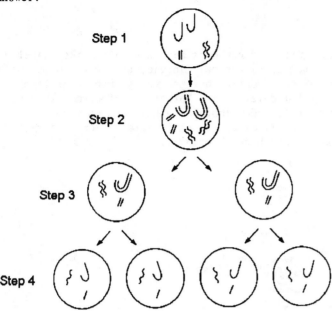

Step 1

Step 2

Step 3

Step 4

Short Response

6. Answer:
   a. There are only two kinds. <u>Sex cells</u>
   b. They contain 23 pairs of chromosomes. <u>Body cells</u>
   c. They unite to form the next generation. <u>Sex cells</u>
   d. They contain 23 chromosomes. <u>Sex cells</u>
   e. There are many different kinds. <u>Body cells</u>

Illustrative

7. Answer:

	T	t
T	TT	Tt
t	Tt	tt

Data for Interpretation

8. Answer:
   a. The risk decreases as pregnancy advances.
   b. Sample answer: The risk is greatest during the first 8 weeks of pregnancy. The risk is high because the eyes, ears, brain, and internal organs are beginning to form during this time.

9. Answer:
Sample answers: Exposure to radiation—Affected newborns may have birthmarks, harelip, cleft palate, deafness, or disorders such as sickle cell anemia, cystic fibrosis, or Down's syndrome. Mother eats nutritious foods during her pregnancy—Embryo or fetus receives adequate nutrients and energy for development. Mother is infected with rubella during first four months of pregnancy—Affected newborns may have one of many birth defects, including deafness, congenital heart disease, mental retardation, cataracts and other eye disorders, cerebral palsy, and bone abnormalities.

Difficulty: 3

Illustrative

10. Answer:

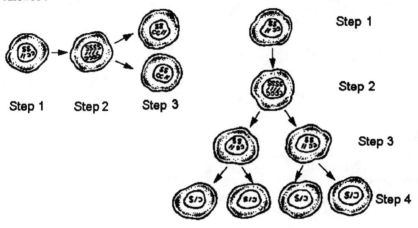

Difficulty: 3

Short Essay

11. Answer:
Sample answer: A flowering cactus illustrates *self-preservation* by keeping other organisms from using its stored water. It illustrates *self-regulation* by carrying out photosynthesis in a dry environment. It illustrates *self-organization* in its ability to repair itself. It illustrates *self-reproduction* in the pollination of its flowers.

Difficulty: 4

12. Answer:
Sample answer: Humans reproduce by sexual reproduction, which produces a new combination of traits in every offspring. By contrast, there are few noticeable differences among the individuals of a species of bacteria because they often reproduce asexually, which results in daughter cells that are identical to the parent cells.

Unit 8 Activity Assessment

Activity Assessment

1. **Pass It On!**
   Teacher's Notes
   Overview
   Students use Punnett squares to find the genotypes that result from hybrid crosses of a fictional organism. They sketch the eight versions of the organism that could arise from their hybrid crosses.

   Materials
   (per activity station)
   · unlined paper
   · a metric ruler
   · a pencil

   Preparation
   Prior to the assessment, equip student activity stations with the materials needed.

   Time Required
   Each student should have 40 minutes to complete the Data Chart.

   Performance
   At the end of the assessment, students should turn in the following:
   · completed Data Chart

   Evaluation
   The following is a recommended breakdown for evaluation of this Activity Assessment:
   · 60% ability to develop accurate Punnett squares
   · 40% ability to interpret data with a sketch

2. **Pass It On!**
   Show how much you know about genetics by creating a fictional organism.

   Before You Begin . . .
   As you work through the tasks, keep in mind that your teacher will be observing the following:
   • how accurately you develop Punnett squares
   • how well you interpret data with a sketch

   Get ready for some genetic engineering!

   Task 1: Of the four different body parts shown below, select three to use in creating your organism.

   Task 2: On a separate piece of paper, draw a Punnett square for each of the three body parts and then cross parents that are both hybrid for that body part. In this activity, all hybrids share the same phenotype as organisms that have two dominant factors. In other words, organisms with the genotypes *BB* and *Bb* look alike. For the genotypes, use the letters shown in parentheses in the chart below.

   Task 3: For the three body parts you chose, eight distinct versions of the organism can result from the hybrid crosses you performed in Task 2. One result for beak, leg, and foot type is shown below at right. Sketch all eight versions of your organism in the Data Chart on the next page.

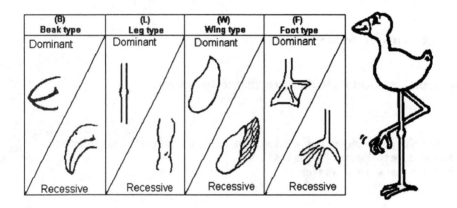

3.

Sketches

Answers to Unit 8 Activity Assessment

Activity Assessment

1. Answer:
   Be sure students understand that the difference between the dominant and recessive forms of a trait are exaggerated in this activity. Fictional organisms are used in order to illustrate a concept.

   Students' Punnett squares should demonstrate that for each hybrid cross, three-fourths of the possible offspring show the dominant trait and one-fourth show the recessive trait.

   Sketches will vary depending on which three body parts each student chose, but the following is a sample answer for the beak, legs, and wings:
   * pointed beak, stick legs, crescent-shaped wing
   * pointed beak, stick legs, notched wing
   * pointed beak, meaty legs, crescent-shaped wing
   * pointed beak, meaty legs, notched wing
   * hooked beak, stick legs, crescent-shaped wing
   * hooked beak, stick legs, notched wing
   * hooked beak, meaty legs, crescent-shaped wing
   * hooked beak, meaty legs, notched wing

2. Answer: Not applicable (student's notes)

3. Answer: See teacher's notes.

Unit 8 SourceBook Assessment

Multiple Choice

1. Differences in height, hair color, or blood type between the parent and offspring of a species are called

   a. variations.   b. mutations.   c. hybrids.   d. factors.

2. When Mendel crossed pure tall pea plants with pure short ones, the offspring were

   a. 100 percent tall.
   b. 50 percent tall and 50 percent short.
   c. 25 percent tall and 75 percent short.
   d. 75 percent tall and 25 percent short.

True/False

3. Mendel's discoveries in the area of heredity were not immediately accepted by the scientific community.

   a. true   b. false

Multiple Choice

4. One factor in a pair that prevents the expression of the other factor is called

   a. dominant.   b. recessive.

5. The process that produces daughter cells which are identical to the parent cell is called

   a. mutation.   b. mitosis.   c. meiosis.   d. fibrosis.

6. The period of cell growth and activity is called

   a. metaphase.   b. telophase.   c. anaphase.   d. interphase.

7. Cell division that results in the production of sex cells is called

   a. mitosis.   b. meiosis.

8. In humans, body cells contain 46 chromosomes. How many chromosomes are there in a human sperm cell?

   a. 92   b. 46   c. 23   d. 12

True/False

9. All living organisms have the same number of chromosomes.

   a. true   b. false

Multiple Choice

10. The factors responsible for the traits of an organism were given the name _____ in 1906.

    a. spindles   b. centromeres   c. genes   d. chromosomes

11. In pea plants, the gene for round seeds is dominant over the gene for wrinkled seeds. *R* is the symbol used for the dominant gene. What symbol would be used to represent the recessive gene?

    a. r   b. W   c. w   d. s

12. When you refer to a friend's hair color you are referring to a

    a. phenotype.   b. genotype.

13. What percentage of the offspring of a cross between a pea plant that is homozygous for yellow seeds and one that is homozygous for green seeds would you expect to have yellow seeds? Yellow seed color is a dominant trait.

    a. 25 percent   b. 50 percent   c. 75 percent   d. 100 percent

14. Sex-linked traits tend to occur more often in

    a. males.   b. females.

15. When the four nitrogen bases in a DNA molecule combine, adenine always pairs with

    a. adenine.   b. thymine.   c. guanine.   d. cytosine.

16. Messenger RNA (mRNA) and transfer RNA (tRNA) work together in the cytoplasm of a cell in order to produce

    a. proteins.   b. enzymes.   c. codons.   d. uracil.

17. What is the name of the molecule of RNA that carries the DNA instructions into the cytoplasm?

    a. codon RNA   b. messenger RNA
    c. anticodon RNA   d. transfer RNA

True/False

18. A mutation is a change in the genetic code that always harms the organism.

    a. true   b. false

Multiple Choice

19. The most serious type of mutation is a

    a. gene mutation.   b. chromosome mutation.

Short Response

20. Using the Punnett square, show a cross between a heterozygous tall pea plant and a homozygous short pea plant.

What percentage of the offspring would you expect to be short?

_____

	T	t
t	Tt	tt
t	Tt	tt

21. Briefly describe the process of DNA replication.

_____

_____

_____

_____

_____

22. Arrange the following in order from smallest to largest: *nucleus, chromosome, cell,* and *gene.*

_____

Matching

23. Match the descriptions on the left with the stages of mitosis on the right.

_____ Nuclear membrane forms, chromosomes become thinner, and cytoplasm divides.

_____ Chromosomes line up along the middle of the cell.

_____ Chromosomes thicken, and spindle fibers form.

_____ Spindle fibers pull duplicate chromosomes to opposite ends of the cell.

a. anaphase

b. metaphase

c. telophase

d. prophase

Short Response

24. Discuss how knowing your family's pedigree with respect to heart disease could benefit you.

_____

_____

_____

_____

_____

_____

_____

_____

25. Give an example of how scientists use recombinant DNA.

_____

_____

_____

_____

_____

Answers to Unit 8 SourceBook Assessment

Multiple Choice

1. Answer: a. variations.

2. Answer: a. 100 percent tall.

True/False

3. Answer: a. true

Multiple Choice

4. Answer: a. dominant.

5. Answer: b. mitosis.

6. Answer: d. interphase.

7. Answer: b. meiosis.

8. Answer: c. 23

True/False

9. Answer: b. false

Multiple Choice

10. Answer: c. genes

11. Answer: c. w

12. Answer: a. phenotype.

13. Answer: d. 100 percent

14. Answer: a. males.

15. Answer: b. thymine.

16. Answer: a. proteins.

17. Answer: b. messenger RNA

True/False

18. Answer: b. false

Multiple Choice

19. Answer: b. chromosome mutation.

Short Response

20. Answer: Fifty percent of the offspring should be short.

21. Answer:
Sample answer: A strand of DNA splits, or unzips, through the action of an enzyme. Freely floating nucleotides attach to their respective bases on the unzipped DNA strand, resulting in the formation of two identical DNA strands.

22. Answer: Gene, chromosome, nucleus, cell

Matching

23. Answer:
    c__ Nuclear membrane forms, chromosomes become thinner, and cytoplasm divides.
    b__ Chromosomes line up along the middle of the cell.
    d__ Chromosomes thicken, and spindle fibers form.
    a__ Spindle fibers pull duplicate chromosomes to opposite ends of the cell.

Short Response

24. Answer:
Sample answer: Because heart disease is known to be affected both by genetics and the environment (such as stress and eating habits), knowing your family's history of heart disease might help you make better decisions about your diet and exercise regimen.

25. Answer:
Answers will vary but may include the production of insulin for use by diabetics, the improvement of food crops, or the treatment of genetic diseases in humans.

Unit 8 Extra Assessment Items

Word Usage

1. Use the following words in two or three sentences to show how they are related: *mitosis, meiosis, cell division, nucleus,* and *sex cells.*

   _____

   _____

   _____

   _____

   _____

2. Use the following words in two or three sentences to show how they are related: *chromosomes, DNA, instructions, linear programming,* and *genes.*

   _____

   _____

   _____

   _____

   _____

Correction/Completion

3. J.T. was asked to write a story illustrating the four essential properties of all living things. Correct the italicized words in J.T.'s story.

   When a gazelle runs from a lion, the gazelle and the lion are both demonstrating (a) *self-regulation.* Both the gazelle and the lion have offspring—a characteristic known as (b) *self-preservation.* The lion pants to cool itself after it has run down its prey. This is called (c) *self-organization.* The food energy and raw materials that the lion obtains from the gazelle are used for the lion's growth, the replacement of its cells, and the repair of its tissues, each of which is an example of (d) *self-reproduction.*

   a. _____

   b. _____

   c. _____

   d. _____

4. Complete the following statements:

   a. Units of genetic instructions are called _____.

      They are segments of _____, which contain DNA. The

      instructions in DNA are "written" in a form of

      _____.

   b. The _____ of a cell controls the formation of new

      cells. Through the process of mitosis, the _____

      in the nucleus are duplicated and then divided between two cells.

      This way, each cell receives an identical set of chromosomes.

   c. The kind of reproduction that involves only one parent and occurs

      by the processes of mitosis and cell division is called

      _____ reproduction. Another form, called

      _____ reproduction, involves two parents and

      involves the union of sex cells that were produced by the process

      of _____.

5. Gregor Mendel grew pea plants and studied how their characteristics were passed on to successive generations. Complete the statements below using the following words: *dominant, factor(s), generation, hybrid, one, purebred, recessive,* and *three.*

   a. For each characteristic, there seemed to be two important

      _____.

   b. In hybrids, one _____ always seemed to be

      stronger and was called _____, while one seemed

      weaker and was called _____.

   c. In _____ plants, the two factors were the same.

   d. A recessive factor could reappear in a later

      _____, even though it was not seen in the

      _____ parents.

   e. There was a definite mathematical relationship of

      _____ to _____, in the

      appearance of the _____ factor to the appearance

      of the _____ factor in the second-generation

      plants.

6. Complete the following statements with *early*, *midway*, or *late* to show approximately when the events of development occurred.

a. Your facial features developed _____ in pregnancy.

b. Your wrinkles were smoothed out _____ in pregnancy.

c. Your bones and muscles formed _____ in pregnancy.

d. You began to move _____ in pregnancy.

e. Your memory began _____ in pregnancy.

f. Your umbilical cord and placenta formed _____ in

pregnancy.

Short Essay

7. *Divide to Multiply* and *One Plus One Equals One* could be book titles that refer to methods of reproduction. Explain why this is so, telling which method of reproduction each title might represent.

_____

_____

_____

_____

_____

_____

Short Response

8. A drug used by pregnant women early in their pregnancy was later linked to birth defects in a small percentage of children. What kinds of birth defects were these children at risk of having? (Hint: Consider which parts of the human embryo form the earliest.)

_____

_____

_____

_____

_____

9. Name and describe the process that produces body cells for growth.

_____

_____

_____

_____

Why would this process be unsuitable for the production of sex cells (ova or sperm)?

_____

_____

_____

_____

_____

_____

_____

10. Name the process by which sex cells are produced, and explain how it works.

_____

_____

_____

_____

_____

11. Imagine that you crossed pea plants with green seeds and pea plants with yellow seeds and that all of the plants produced had yellow seeds.

a. Describe how you would proceed to get pea plants with green seeds again.

_____

_____

_____

_____

_____

b. Describe the laws of heredity that are involved in the process you described in (a). In your description, use the following words: *hybrid, dominant, first generation,* and *recessive.*

_____

_____

_____

_____

_____

_____

12. What can a woman do to protect her unborn child from being born with birth defects?

_____

_____

_____

13. List four body parts in the order in which they develop in a human embryo.

_____

_____

_____

_____

14. Name two body parts that continue to develop (not just grow larger) after birth.

_____

_____

_____

15. Which do you know more about, your phenotype or your genotype? Explain your answer with examples.

_____

_____

_____

_____

_____

_____

16. Give two examples of linear programming.

_____

_____

_____

_____

17. Explain why chromosomes can be called a form of linear programming.

_____

_____

_____

_____

Graphic

18. Examine the data presented below.

Incidence of Diabetes in Offspring of Diabetic Mother or Father

Age of offspring (in years)	Only mother diabetic (percent)	Only father diabetic (percent)
30	2.1	2.0
40	3.6	3.2
50	5.8	8.0
60	9.6	13.2
70	11.3	11.7

a. Construct a graph using this data to show how having a diabetic mother or father affects the offspring's chances of developing diabetes.

b. Does this graph support the view that diabetes is influenced by genetic factors?

_____

_____

_____

_____

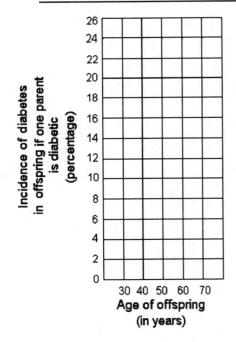

19. In rabbits, black coat color (*B*) is dominant over brown coat color (*b*). Straight hair (*S*) is dominant over curly hair (*s*). Someone gives you a female rabbit with straight black hair and a male rabbit with curly brown hair.

a. What are the possible genotypes for the male?

_____

b. What are the possible genotypes for the female?

_____

c. If you bred these two, would you expect to get any rabbits with curly brown hair in the first generation?

_____

_____

_____

_____

Explain your answer with Punnett squares. Use labels to indicate any combination that yields curly brown hair.

20. a. Use the data presented below to construct a bar graph that shows the likelihood (in percentages) that if one twin has a certain disorder, the other one will have the same disorder.

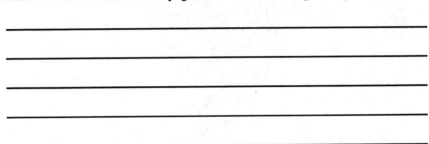

Disorder	Identical twins	Fraternal twins
diabetes	55%	10%
allergies	50%	4%
overactive thyroid	47%	3%
gallstones	27%	6%

b. In your opinion, does the information presented here provide evidence that these disorders are influenced by genetic factors? Explain your answer.

_____

_____

_____

_____

_____

Incidence of Both Twins Having Certain Disorders

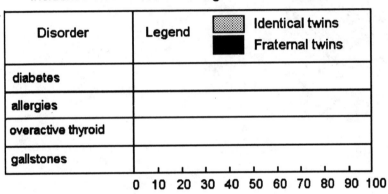

HRW material copyrighted under notice appearing earlier in this work.

389

Illustrative

21. For each of the following illustrations, what essential property of living things is being illustrated?

a. _____

b. _____

c. _____

22. Use the information given in the following birth announcement to answer the questions that follow.

I'm Cecily Catherine Chase. I was born to my parents, Craig and Irene, at 1:30 A.M. on June 29, 1996, at General Hospital. I have two brothers, Erik and Ian, and two sets of grandparents, John and Cathy Chase and David and Sonia Dawe. I'm looking forward to meeting Aunt Betty and Uncle Gregory Chase, their kids Mark and Emma Chase, and Uncle Nick Dawe.

a. Construct a family tree for the above information.

b. Use the family tree you constructed above to answer the following questions: How is Cecily related to Mark and Emma?

_____

How is Craig related to
• John?

_____

• Emma?

_____

• Nick?

_____

• Cecily?

_____

c. List the people from whom Cecily might inherit her genes.

_____

_____

_____

_____

d. Name someone who does not appear to have any ancestors in common with Cecily.

_____

23. a. Study the pedigree below by tracing the trait of dimples through three generations of a family. Using the information given in the key, label the genotypes for each person in the family. Dimples are a dominant trait, so use *D* to show the dominant gene (dimples) and *d* to show the recessive gene (no dimples).

b. For each of the two marriages in the family, draw a Punnett square to show the possible gene combinations for the trait of dimples in the offspring.

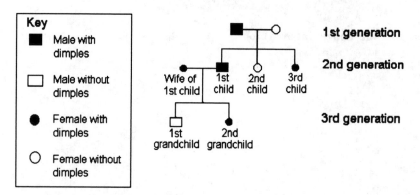

Key
- ■ Male with dimples
- □ Male without dimples
- ● Female with dimples
- ○ Female without dimples

1st generation

2nd generation

Wife of 1st child | 1st child | 2nd child | 3rd child

3rd generation

1st grandchild | 2nd grandchild

24. Answer the following questions about two black (*BB* or *Bb*) rabbits.

a. Use a Punnett square to show the results of mating two black rabbits, both of which have the genes *Bb* for color.

b. If 12 baby rabbits are born, how many might you expect to be black?

_____

25. Kaylee gathered the information below about members of her family. Use it to make a family pedigree for Kaylee to show the inheritance of the dominant gene for dimples.

Family Member	Dimples
Kaylee	yes
Mom (Ann)	no
Dad (Barry)	yes
Brother (Len)	yes
Sister (Beth)	no
Dad's mother (Helen)	yes
Dad's father (Joe)	no
Dad's sister (Millie)	yes

26. Suppose that you want to find out whether Fluffums, your beautiful black cat, carries the recessive gene for gray. Which of the following cats would be most suitable as a mate to determine this characteristic? Explain. (*D* represents the dominant gene for a black coat and *d* represents the recessive gene for a gray coat).
   a. Ebony, a black cat (*DD*)
   b. Boots, a black cat (*Dd*)
   c. Kramer, a gray cat (*dd*)
   Use Punnett squares to prove that you have chosen the best mate for this purpose. Calculate the percentage of kittens you would expect to be black and the percentage you would expect to be gray. (If there are no gray kittens out of a litter of at least 7, the parent cat is considered to be pure black, *DD*).

   _____

   _____

   _____

   _____

   _____

Numerical Problem

27. The specific gene that causes a disorder called myotonic distrophy has been identified. About 1 in 8000 people have this disorder. How many people in the United States probably have the disorder? (Base your answer on a population of 249,600,000)

   _____

Performance Task

28. **Materials:** clay, pipe cleaners, toothpicks, markers

   Task: Prepare a set of models that could be used to illustrate what happens to chromosomes during (a) mitosis and (b) meiosis.

Answers to Unit 8 Extra Assessment Items

Word Usage

1. Answer:
   Sample answer: New body cells are made by a type of *cell division* called *mitosis*. During mitosis, the material in the *nucleus* is doubled and then divided into two equal parts. *Sex cells* are made by *meiosis*, the process by which the material in the nucleus is doubled and then divided twice.

2. Answer:
   Sample answer: *Genes* are the units of genetic *instructions*. They are segments of *chromosomes*, which contain *DNA*. The instructions in DNA are "written" in a form of *linear programming*.

Correction/Completion

3. Answer:
   a. Self-preservation
   b. Self-reproduction
   c. Self-regulation
   d. Self-organization

4. Answer:
   a. Units of genetic instructions are called <u>genes</u>. They are segments of <u>chromosomes</u>, which contain DNA. The instructions in DNA are "written" in a form of <u>linear programming</u>.
   b. The <u>nucleus</u> of a cell controls the formation of new cells. Through the process of mitosis, the <u>chromosomes</u> in the nucleus are duplicated and then divided between two cells. This way, each cell receives an identical set of chromosomes.
   c. The kind of reproduction that involves only one parent and occurs by the processes of mitosis and cell division is called <u>asexual</u> reproduction. Another form, called <u>sexual</u> reproduction, involves two parents and involves the union of sex cells that were produced by the process of <u>meiosis</u>.

5. Answer:
   a. For each characteristic, there seemed to be two important <u>factors</u>.
   b. In hybrids, one <u>factor</u> always seemed to be stronger and was called <u>dominant</u>, while one seemed weaker and was called <u>recessive</u>.
   c. In <u>purebred</u> plants, the two factors were the same.
   d. A recessive factor could reappear in a later <u>generation</u>, even though it was not seen in the <u>hybrid</u> parents.
   e. There was a definite mathematical relationship of <u>three</u> to <u>one</u>, in the appearance of the <u>dominant</u> factor to the appearance of the <u>recessive</u> factor in the second-generation plants.

6. Answer:
   a. Your facial features developed <u>midway</u> in pregnancy.
   b. Your wrinkles were smoothed out <u>late</u> in pregnancy.
   c. Your bones and muscles formed <u>early</u> in pregnancy.
   d. You began to move <u>early</u> in pregnancy.
   e. Your memory began <u>late</u> in pregnancy.
   f. Your umbilical cord and placenta formed <u>early</u> in pregnancy.

Short Essay

7. Answer:
The first title refers to asexual reproduction, or mitosis. Through mitosis, a cell divides to form two identical cells. The second title refers to sexual reproduction, or meiosis. Two sex cells formed through meiosis can unite to form a single cell.

Short Response

8. Answer:
Sample answer: The basic body plan, organs, and organ systems develop during the first trimester of pregnancy, so birth defects of the heart, brain, spine, digestive tract, limbs, lungs, liver, kidneys, bones, or muscles are possible.

9. Answer:
Growth occurs through mitosis, a process that takes place in the nucleus of a cell. The result is a division of the cell into two cells, each with the same number of chromosomes as the original cell.

Sex cells must have one-half of the number of chromosomes found in the cell that results from the union of a male sex cell and female sex cell. In mitosis, the cells that result from division of the original cell are each identical to their parent. In other words, they have a complete set of chromosomes, rather than the one-half required by sex cells.

10. Answer:
During meiosis, the nucleus of a cell divides to produce two sex cells, each having only half the number of chromosomes found in the parent cell. A female sex cell and a male sex cell can therefore combine to form a new cell that is different from the cells found in either parent.

11. Answer:
    a. The yellow seeds are hybrids. Cross-pollinate the plants with yellow seeds, and plant the resulting seeds. One-quarter of the resulting plants should have green seeds
    b. Cross-breeding produces *hybrids* in the *first-generation* offspring. Each of these offspring has the *dominant* trait. One-quarter of the offspring in the second generation, however, have the *recessive* trait.

12. Answer:
To protect her unborn child, a woman should abstain from smoking, taking drugs, and drinking alcoholic beverages.

13. Answer:
There are many possible answers. If in doubt about the answer given, check it against the information in Adventures of a Life in Progress, which begins on page 519 of the textbook.

14. Answer: Brain and bone structure

15. Answer:
Sample answer: A person's phenotype consists of his or her observed inherited traits, while genotype relates to a person's genetic makeup. A person should therefore know more about his or her phenotype than about his or her genotype. For example, a person can observe that their phenotype includes brown eyes. Their genotype, however, may actually include a "brown eye" gene and a "blue eye" gene.

16. Answer:
Examples of linear programming involve a series of symbols or characters arranged in rows or lines, such as this sentence, a bar code, or a computer program.

17. Answer:
Sample answer: Chromosomes contain long strands of DNA. The DNA is made up of four chemical "letters" that can be arranged in various combinations, with each combination producing a different result. Thus, they are a form of linear programming.

Graphic

18. Answer:
   a. See graph below.
   b. The graph shows that the chances of developing diabetes if one parent has the disease are much greater than for those whose parents are both free of the disease.

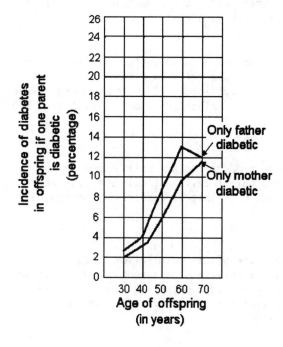

19. Answer:
    a. bb and ss
    b. BB, Bb, SS, and Ss
    c. The only way for the first-generation offspring to have curly, brown hair would be
       if the female rabbit was hybrid for both characteristics. See diagrams below. Students
       should draw at least the top two Punnett squares, but they may have also drawn the
       bottom two in the process of finding an answer.

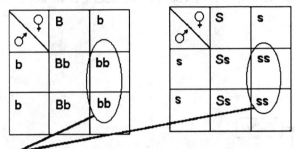

A combination of these genotypes will produce brown curly hair.

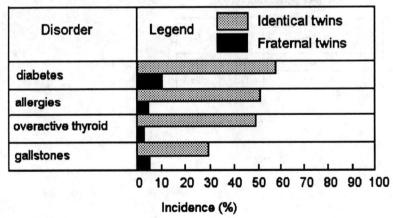

20. Answer:
    a. See graph below.
    b. Answers will vary. Students should point out how much greater the chances are of
       having the same disorder if twins are identical. Identical twins are genetically the
       same, while fraternal twins are not, so there is evidence in the information.

Incidence of Both Twins Having Certain Disorders

Disorder	Legend	Identical twins
		Fraternal twins
diabetes		
allergies		
overactive thyroid		
gallstones		

0 10 20 30 40 50 60 70 80 90 100

Incidence (%)

Illustrative

21. Answer:
    a. Self-organization
    b. Self-reproduction
    c. Self-preservation

22. Answer:
   a. See diagram below.
   b. They are cousins.
      Craig is John's son (or John is Craig's father).
      Craig is Emma's uncle (or Emma is Craig's niece).
      Craig is Nick's brother-in-law (or Nick is Craig's brother-in-law).
      Craig is Cecily's father (or Cecily is Craig's daughter).
   c. Her parents (Irene and Craig Chase) and her grandparents (David and Sonia Dawe and John and Cathy Chase)
   d. Betty

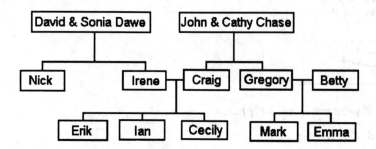

23. Answer:

   a.

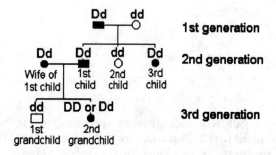

   b.

First-generation cross		
Female Genes / Male Genes	d	d
D	Dd	Dd
d	dd	dd

Second-generation cross		
Female Genes / Male Genes	D	d
D	DD	Dd
d	Dd	dd

24. Answer:
    a. See diagram below.
    b. You would expect 9 of the 12 offspring to be black.

♂\♀	B	b
B	**BB**	**Bb**
b	**Bb**	**bb**

25. Answer: Family pedigree for Kaylee:

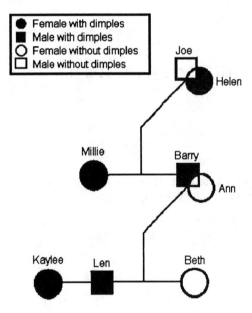

- ● Female with dimples
- ■ Male with dimples
- ○ Female without dimples
- □ Male without dimples

26. Answer:
All of Ebony's offspring would be black, so he is not the right choice. One-quarter of the offspring of Boots and a black cat with a recessive gene for gray would be expected to be gray. However, in a litter with only seven cats, the absence of a gray offspring would not be decisive evidence that Boots does not have a recessive gene for gray. Kramer is the best choice. As shown below, half of the offspring would be expected to be gray if Fluffums has a recessive gene for gray, whereas there would be no gray offspring if Fluffums does not have a recessive gene for gray.

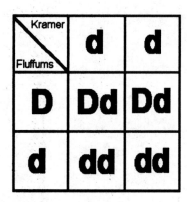 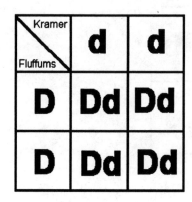

Numerical Problem

27. Answer:
Based on a population of 249,600,000, there are probably 31,200 people with this disorder in the United States.

Performance Task

28. Answer:
Answers will vary but should show the logical use of materials to effectively demonstrate what happens to chromosomes during mitosis and meiosis.